AF790377

IntechOpen

Supporting open minds since 2005

Advanced Oxidation Processes - Applications, Trends, and Prospects
http://dx.doi.org/10.5772/intechopen.85681
Edited by Ciro Bustillo-Lecompte

Contributors
Ciro Fernando Bustillo-Lecompte, Gianluca Li Puma, Fiderman Machuca-Martínez, Miguel Mueses, José Colina-Márquez, Changyong Wu, Yu Tan, Xiumei Sun, Yanan Li, Yuexi Zhou, Liya Fu, Min Xu, Ting Zhang, Prabir Ghosh, Damodhar Ghime, Milan Malhotra, Lakshmi Pisharody, Ansaf V Karim, Sukanya Krishnan, Zhenhua Zhang, Li-Ping Fan, Yue-Juan Wang, Lars Rehmann, Zhaoran Xin

© The Editor(s) and the Author(s) 2020
The rights of the editor(s) and the author(s) have been asserted in accordance with the Copyright, Designs and Patents Act 1988. All rights to the book as a whole are reserved by INTECHOPEN LIMITED. The book as a whole (compilation) cannot be reproduced, distributed or used for commercial or non-commercial purposes without INTECHOPEN LIMITED's written permission. Enquiries concerning the use of the book should be directed to INTECHOPEN LIMITED rights and permissions department (permissions@intechopen.com).
Violations are liable to prosecution under the governing Copyright Law.

BY-NC

Individual chapters of this publication are distributed under the terms of the Creative Commons Attribution - NonCommercial 4.0 International which permits use, distribution and reproduction of the individual chapters for non-commercial purposes, provided the original author(s) and source publication are appropriately acknowledged. More details and guidelines concerning content reuse and adaptation can be found at http://www.intechopen.com/copyright-policy.html.

Notice
Statements and opinions expressed in the chapters are these of the individual contributors and not necessarily those of the editors or publisher. No responsibility is accepted for the accuracy of information contained in the published chapters. The publisher assumes no responsibility for any damage or injury to persons or property arising out of the use of any materials, instructions, methods or ideas contained in the book.

First published in London, United Kingdom, 2020 by IntechOpen
IntechOpen is the global imprint of INTECHOPEN LIMITED, registered in England and Wales, registration number: 11086078, 7th floor, 10 Lower Thames Street, London, EC3R 6AF, United Kingdom
Printed in Croatia

British Library Cataloguing-in-Publication Data
A catalogue record for this book is available from the British Library

Additional hard and PDF copies can be obtained from orders@intechopen.com

Advanced Oxidation Processes - Applications, Trends, and Prospects
Edited by Ciro Bustillo-Lecompte
p. cm.
Print ISBN 978-1-78984-890-8
Online ISBN 978-1-78984-891-5
eBook (PDF) ISBN 978-1-83880-971-3

An electronic version of this book is freely available, thanks to the support of libraries working with Knowledge Unlatched. KU is a collaborative initiative designed to make high quality books Open Access for the public good. More information about the initiative and links to the Open Access version can be found at www.knowledgeunlatched.org

For EU product safety concerns:
IN TECH d.o.o., Prolaz Marije Krucifikse Kozulić 3, 51000 Rijeka, Croatia, info@intechopen.com or visit our website at intechopen.com.

We are IntechOpen, the world's leading publisher of Open Access books

Built by scientists, for scientists

4,800+
Open access books available

123,000+
International authors and editors

140M+
Downloads

151
Countries delivered to

Our authors are among the
Top 1%
most cited scientists

12.2%
Contributors from top 500 universities

CLARIVATE ANALYTICS
BOOK CITATION INDEX
INDEXED

WEB OF SCIENCE™

Selection of our books indexed in the Book Citation Index
in Web of Science™ Core Collection (BKCI)

Interested in publishing with us?
Contact book.department@intechopen.com

Numbers displayed above are based on latest data collected.
For more information visit www.intechopen.com

Meet the editor

Dr. Ciro Bustillo LeCompte has a multidisciplinary background in the areas of civil, environmental, and chemical engineering. He completed his Bachelor of Engineering (2008) at the University of Cartagena, Colombia, and obtained his MASc (2012) and PhD (2016) at Ryerson University, Canada. Dr. LeCompte is a certified Professional Engineer (PEng), Environmental Professional (EP), a Fraternal Member of the Canadian Institute of Public Health Inspectors (CIPHI), and a 2017-2018 Queen Elizabeth Scholar (QES). He is currently an Associate Member in the Environmental Applied Science and Management Graduate Programs, and a Lecturer in the School of Occupational and Public Health at Ryerson University. He has co-authored over twenty peer-reviewed scientific papers, several conference proceedings, chapters, and books. His research interests include advanced oxidation processes, advanced treatment of water and wastewater, waste minimization, water, soil and air quality, energy and resource recovery, water reuse, and heterogeneous catalysis.

Contents

Preface

Advanced Oxidation Processes - Applications, Trends, and Prospects constitutes a comprehensive resource for civil, chemical, and environmental engineers researching in the field of water and wastewater treatment. The book covers the fundamentals, applications, and future work in Advanced Oxidation Processes (AOPs) as an attractive alternative and a complementary treatment option to conventional methods.

This book presents state-of-the-art research on AOPs and heterogeneous catalysis while covering recent progress and trends, including the application of AOPs at the laboratory, pilot, or industrial scale, the combination of AOPs with other technologies, hybrid processes, process intensification, reactor design, scale-up, and optimization. The book is divided into four sections: Introduction to Advanced Oxidation Processes, General Concepts of Heterogeneous Catalysis, Fenton and Ferrate in Wastewater Treatment, and Industrial Applications, Trends, and Prospects.

The first section provides the reader with a comprehensive introduction to AOPs, organic contaminants, current technologies, heterogeneous catalysis, mechanisms, and application in wastewater treatment. Fundamentals and general concepts of heterogeneous catalysis are the main focus of the second section, with particular emphasis on chemical kinetics and its application for wastewater treatment. Section 3 is focused on the use of Fenton and Ferrate as wastewater treatment methods; particular attention is given to fundamentals and applications of Fenton and Ferrate, followed by future research, guidelines on regeneration, and environmental impact. Finally, the book closes with a section on the application of AOPs in different industries, such as food and beverages, petrochemical, and flexographic printing; future trends and prospects are described, as well as a comprehensive case study of a solar photocatalytic plant, where the scale-up and optimization processes were covered from bench- and pilot-scale to full-scale operation.

I would like to acknowledge the authors who collaborated with their research towards this book; their efforts and contributions are highly appreciated. I would like to thank both Ms. Maja Bozicevic and Mr. Edi Lipović, Author Service Managers, for their support during the editing and publishing process, as well as the partnership of IntechOpen and Knowledge Unlatched for fully funding and sponsoring this endeavour.

To my wife, Liseth, my daughter, Victoria, and my entire family, with love.

Dr. Ciro Bustillo-Lecompte
School of Occupational and Public Health,
Graduate Programs in Environmental Applied Science and Management,
Ryerson University,
Toronto, Canada

Section 1

Introduction to Advanced Oxidation Processes

Chapter 1

Advanced Oxidation Processes: A Powerful Treatment Option for the Removal of Recalcitrant Organic Compounds

Damodhar Ghime and Prabir Ghosh

Abstract

Advanced oxidation processes (AOPs) are the technologies that generally use the hydroxyl radicals, the ultimate oxidant for the remediation of organic contaminants in wastewater. These are highly effective novel methods speeding up the oxidation process. AOP can combine with ozone (O_3), catalyst, or ultraviolet (UV) irradiation to offer a powerful treatment of wastewater. Future research should be focused on enhancing the properties of heterogeneous catalysts in AOPs. This chapter reports general review of different AOPs utilized for the removal of various phenolic compounds and textile dyes in wastewater. The chapter also aimed at an investigation of efficiency for different photochemical AOPs. The authors have carried out the experimental runs at a laboratory scale for the removal of malachite green oxalate (MGO) dye with photochemical AOPs. The influence of ferrous ions and oxidant dosage on percentage decolorization of MGO in wastewater has been reported. The discussion extends to the utilization of different modified photocatalysts for the photocatalysis process. The future challenges, such as the adoption of strategies for the integration of processes and the decrement in operational cost of AOPs, are discussed. The discussion covers the utilization of different heterogeneous catalysts, the reduction of input demands of chemicals and energy for the processes.

Keywords: advanced oxidation processes (AOPs), wastewater treatment, organic contaminants, heterogeneous catalysts, hydroxyl, sulphate radicals

1. Introduction

Advanced oxidation processes (AOPs) are the most attractive and favorable option for the effective removal of organic pollutants in wastewater. In recent years, many harmful chemicals in the released effluent from industries have been identified in the environment. These chemicals are hazardous to both human beings and aquatic biota. They can cause severe damages to human beings and the marine environment. So, there is a strong need to treat toxic pollutants with a proper treatment option. Thus, AOPs find its suitable applicability for the treatment of wastewater containing harmful chemicals. Advanced oxidation technologies are based on the in situ generations of strong oxidants, i.e. hydroxyl radicals and sulphate radicals, for the oxidation of organic pollutants [1]. Some of the AOPs use ozone and UV irradiation for better treatment efficiency. These AOP technologies have already been established and started at full scale for the treatment of drinking water and the facilities of water reuse. Also, several AOPs such as electrochemical treatment, use of electron beam, plasma, microwave and ultrasound related processes are continually being studied by the environmental researchers and scientists.

AOPs are very well known for bridging the gap between the treatability obtained by traditional physicochemical and biological processes, and from day to day more stringent regulations are set by the environmental laws. As extensive research is going on in AOPs from the last two decades, still these technologies are well matured, and there are again some fields worthy of research [2].

1.1 Different AOPs for wastewater remediation

Different AOPs such as photocatalysis, Fenton-like processes and ozonation were investigated for the treatment of an integrated solution of produced water that contains toluene, xylene, naphthalene, phenol, malonic and acetic acids. This integrated solution of produced water was referred as a seawater matrix [3]. The efficiency in terms of total organic carbon (TOC) was studied. Among different applied AOPs, photocatalysis was less efficient for the treatment of a seawater matrix. The removals of TOC lower than 20% were noticed in 4 h of treatment. The better results were obtained in the ozonation process that combined with oxidant hydrogen peroxide (H_2O_2). The combined effect of H_2O_2 and ozone leads to the complete removal of organics, consisting of a high-percentage removal of acetic acid that was not entirely oxidized with the rest of AOPs used. 74% TOC removal efficiency was observed in ozonation for the optimum conditions of 4 g/h ozone, 500 mg/L H_2O_2 and pH of 10 in 2 h of treatment time.

Several AOPs such as O_3, O_3/H_2O_2, UV, UV/O_3, UV/H_2O_2, O_3/UV/H_2O_2, Fe^{2+}/H_2O_2 and photocatalysis processes were investigated for the oxidation of phenol in aqueous medium [4]. The comparison of these different AOPs was undertaken considering some of the experimental parameters. Among all, the Fenton process showed the fastest removal rate for phenol in wastewater. The lower costs were observed for ozonation. Single ozonation provides the best results for phenol degradation in ozone combinations.

The viability of the ozonation and ozone combined with UV photolysis processes was investigated for the degradation of direct blue 86 dye in wastewater [5]. The batch experiments were conducted for studying the effect of solution pH, initial dye concentration and reaction time. The obtained results show that the pH and initial dye concentration controls the efficiency of the process. More than 98% decolorization efficiency was obtained at pH 11, 100 mg/L of initial dye concentration in 35 min of ozone treatment.

All AOPs are characterized by their potential of exploiting the higher reactivity of hydroxyl radicals while driving in oxidation processes. The different AOPs with their essential features are discussed and presented for the wastewater remediation [6]. The number of investigations was carried out with the Fenton process for the abatement of organic pollutants in wastewater. Fenton's reagent can destroy the toxic contaminants in the wastewater like phenols and pesticides. The use of Fenton's reagent for the oxidation purpose is an attractive option for the wastewater treatment due to the iron abundances, and hydrogen peroxide is more comfortable to handle and environmentally safe. Sometimes, UV irradiation is also used with the Fenton process to increase the process efficiency. This process can overcome the shortcomings of the Fenton process. Photolysis allows the regeneration of ferrous ions (Fe^{2+}) from ferric ions (Fe^{3+}) that produce sludge in the Fenton process. Besides, the titanium dioxide (TiO_2) in anatase form finds its application in photocatalysis at a greater level due to its higher stability, low cost and better performance for the degradation of organics.

1.2 Organic pollutants in industrial wastewater

The contamination of water resources continuously exists due to the release of organic pollutants from industrial discharges. The presence of organic pollutants affects the freshwater quality and causes severe effects on aquatic biota. The recent improvement in very sensitive analytical techniques is able to measure the concentration of organic pollutants, even in μg/L concentrations. The organic content generally measures in terms of total organic carbon, chemical oxygen demand (COD), and biological oxygen demand (BOD) [7]. In most of the cases, the dyes and phenolic compounds were taken as model pollutants as per the published literature in recent years. Among different organic pollutants, the phenolic compounds have deserved more attention owing to their high toxicity and frequency of industrial wastewaters [8]. The presence of phenolic compounds in wastewater gives unpleasant smell and taste even at deficient concentrations.

Agro-industrial wastewaters are characterized by the presence of a higher amount of organic and inorganic pollutants of environmental concern [9]. Some parameters, such as high pollutant dose and production of more volume, make the treatment of agro-industrial waters an ecological challenge. For the effective treatment of industrial effluents, the wastewater treatment should consist of degradation as well as the mineralization of water pollutants. During oxidation, the conversion of identified pollutants in wastewater to its more stable oxidation state is necessary.

1.3 Aim and objectives of the chapter

This chapter studies the general review of different AOPs employed for the oxidation of organic pollutants in wastewater. The chapter also aimed at an investigation of the efficiency of different photochemical AOPs. Experimental runs using a UV photochemical reactor were performed at a laboratory scale for the decoloration of malachite green oxalate (MGO) dye using photochemical AOPs. The influence of ferrous ions and oxidant dosage on percentage decolorization of MGO in wastewater has been reported. The discussion was also extending to the utilization of different modified photocatalysts for the photocatalysis process. Some of the future challenges, such as the adoption of strategies for the integration of processes and the decrement in operational cost of AOPs, are discussed.

2. Different photochemical AOPs

Photochemical AOPs are frequently used as an efficient barrier for the oxidation of organic pollutants in wastewater. The combined effect of UV irradiation and hydrogen peroxide (H_2O_2) reaction confirms the removal of a broader range of compounds. Common photochemical AOPs used are UV alone, UV/H_2O_2, UV/Fe^{2+}, UV/H_2O_2/Fe^{2+}, UV/O_3, UV/$S_2O_8^{2-}$, UV/TiO_2, UV/chlorine and UV with some other photocatalysts [10]. Therefore, UV AOPs can show better removal rates with both the radicals, i.e. hydroxyl and sulphate radicals.

2.1 Hydroxyl radical-based photochemical AOPs

The removal of targeted micro-pollutants, including naproxen, carbamazepine, diclofenac, gemfibrozil, ibuprofen, caffeine and mecoprop, were reported with photolysis and UV/H_2O_2 processes [11]. The medium-pressure UV lamp was used for the UV irradiation. The obtained results indicate that UV photolysis alone is not useful for the removal of micro-pollutants, excluding diclofenac and mecoprop. So, there was a need to treat the rest of the micro-pollutants with the UV/H_2O_2 process. It was found that the overall rates of oxidation for all selected organic pollutants were enhanced in the presence of H_2O_2, especially in the case of caffeine and carbamazepine at an oxidant dose of 25 mg/L on comparing with direct UV photocatalysis. Accordingly, the addition of H_2O_2 is critical in the photolysis process for better removal efficiency as it can promote the degradation of micro-pollutants in wastewater. The radical species are generated through the transfer of electrons by transition metal oxidation of oxidants or by the photolysis of H_2O_2 or thermolysis of H_2O_2 and in heterogeneous TiO_2 photocatalysis [12].

Recently, UV photolysis of hydrogen peroxides has earned considerable significance as they are comparatively a benign process concerning the environmental and economic considerations. The activation of symmetrical peroxides, mainly H_2O_2 with UV irradiation, generates the hydroxyl radical species ($^{\bullet}OH$) through the homolytic cleavage of the peroxide bond. These hydroxyl radicals are responsible for the oxidation of water pollutants.

2.2 Sulphate radical-based photochemical AOPs

Sulphate radical-based photochemical AOPs have drawn considerable attention because of the typical advantages of sulphate radicals such as higher redox potential (2.5–3.1 V), stability, selectivity and its use on the broader pH ranges. These radicals could be generated from different persulphates under the application of light, heat and transition metal and sometimes in an alkaline atmosphere or during electrochemical treatment [13]. The combined effect of UV irradiation with persulphate confirms the destruction of a broader range of water pollutants in wastewater.

The UV-activated persulphate advanced oxidation process was investigated for the degradation of sulfamethoxypyridazine (SMP) [14]. The sulphate radical-based photochemical AOP can significantly remediate SMP from aqueous solutions. In a comparative study of kinetics, the second-order reaction rate constant for SMP with sulphate radicals was obtained to $2.73 \times 10^{10}\ M^{-1}\ s^{-1}$, whereas the use of hydroxyl radicals was $2.22 \times 10^{10}\ M^{-1}\ s^{-1}$. Therefore, the faster reaction kinetics was in the case of sulphate radical-based photochemical AOPs.

Moreover, the uses of different heterogeneous catalysts such as cobalt, silver, copper and ruthenium-based catalysts have been studied to efficiently activate peroxymonosulphate (PMS) and peroxydisulphate (PDS) for the removal of water contaminants. Comparing to other activation methods, the activation of persulphates with heterogeneous catalysts has its own merits, such as the requirement of lesser energy, ease of scale-up and higher reactivity. Nevertheless, the use of cobalt- or copper-based catalysts might create some health problems because of the leaching of harmful metal into the aqueous phase, which can stop their practical applications. Again, the use of ruthenium-based catalysts for persulphate activation can be more costly. Thus, Mn-based catalysts are the best option to use for the activation of persulphate. Mn-based materials show some crucial advantages like its higher natural abundance, lower level of toxicity and environmental friendliness.

So, these materials have paid more attention to further improvement in Mn-based catalysts for the activation of persulphates [15].

3. Decolorization of malachite green oxalate dye through photochemical AOPs

The cylindrical type of photochemical reactor was used for the decolorization of malachite green oxalate dye in wastewater. The technical details for the reactor were that it was having the effective volume of 1.47 L and feeding tank volume of 5 L. The reactor was also built with an inside reflecting surface and four tubes of UV source (low-pressure mercury-vapor lamp of intensity 8 W each) surrounded with glass/quartz tube which was placed centrally in reactor [16]. The feeding tank was made up of SS304 material. The wastewater was circulated into the system with the help of a magnetic pump that was composed of polypropylene material. The maximum working temperature for the photochemical reactor was 80°C. The unknown concentration of MGO dye in wastewater after the treatment was measured with a UV–vis spectrophotometer. Absorbance values were detected at 618 nm of maximum wavelength for MGO dye.

3.1 Comparative assessment of different photochemical AOPs for MGO dye

The comparative assessment of various photochemical AOPs, including UV alone, UV/H_2O_2, UV/TiO_2, UV/H_2O_2/TiO_2 and UV/Fe^{2+}/H_2O_2, were investigated for the decolorization of MGO dye. The reaction conditions used were initial dye and H_2O_2 concentrations of 100 mg/L and 12 mM, pH 3, ferrous ion concentration of 60 mg/L and TiO_2 dose to 0.6 g/L with irradiation of 32 W for 60 min of treatment time. A higher rate of decolorization was found in the case of UV/H_2O_2/TiO_2 and UV/Fe^{2+}/H_2O_2 processes. It was about 97 and 98% for UV/Fe^{2+}/H_2O_2 and UV/H_2O_2/TiO_2 processes, respectively. This was happened due to the higher rate of generation for hydroxyl radicals in both the processes. The use of photocatalyst (i.e. TiO_2) enhances the production of hydroxyl radicals in the process, leading to the higher percentage removal of MGO dye. UV photolysis alone was not able to decolorize the MGO completely (77% for a treatment time of 60 min). UV/H_2O_2 and UV/TiO_2 processes almost achieved 95% decolorization efficiency. In short, the decolorization of MGO dye was reduced in the order of [16]:

$$UV/Fe^{2+}/H_2O_2 > UV/H_2O_2/TiO_2 > UV/TiO_2 > UV/H_2O_2 > UV\ alone \quad (1)$$

3.2 Influence of ferrous ions on MGO dye decolorization in the photo-Fenton process

The influence of ferrous ions on MGO dye decolorization was studied by taking the concentrations of Fe^{2+} in the range of 20–80 mg/L. **Figure 1** shows the effect of varying ferrous ion concentration on MGO dye decolorization in the photo-Fenton process. It was seen that the rate of decolorization was enhanced as the ferrous ion concentrations increased up to 20–60 mg/L. Further increase in ferrous ions to 80 mg/L decreased the rate of decolorization. This might have happened due to the rise in the generation of hydroxyl radicals for a higher concentration of Fe^{2+} ions. It can show some scavenging effects of hydroxyl radicals. More concentration of ferrous ions in solution brings about the brown turbidity of ferric hydroxides, and this can cause the recombination of hydroxyl radicals in the aqueous medium [16].

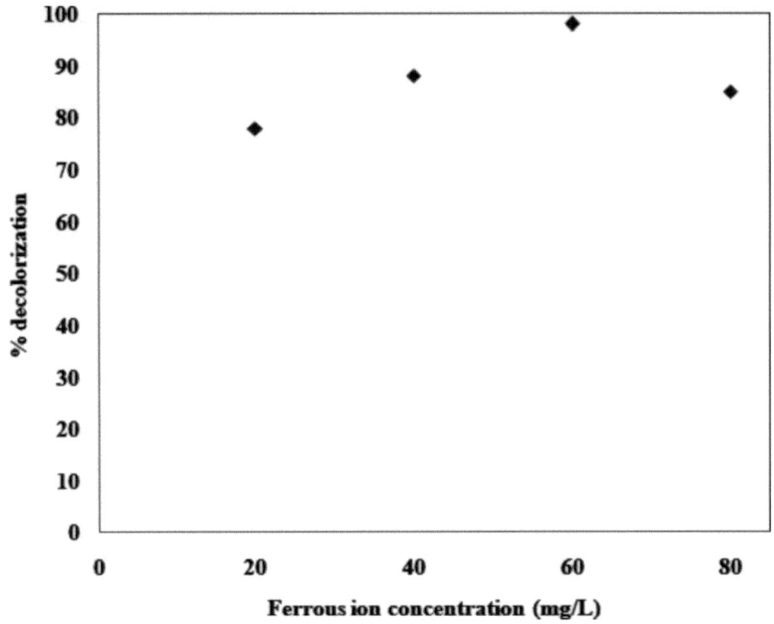

Figure 1.
Influence of ferrous ions (Fe^{2+}) on MGO dye decolorization with the photo-Fenton process by using dye concentration of 100 mg/L, 12 mM of H_2O_2 and pH 3 with irradiation of 32 W for 60 min of treatment time.

3.3 Effect of H_2O_2 concentrations in UV/H_2O_2, photo-Fenton and TiO_2 photocatalysis processes

The concentration of oxidant in the photochemical AOPs (UV/H_2O_2, photo-Fenton and TiO_2 photocatalysis) was varied in the range of 6–36 mM. About 88.61% of decolorization was observed for a dose of 6 mM in the photo-Fenton process [16]. This shows the existence of the generation of hydroxyl radicals into the system. Increment in percentage decolorization was observed with an increase in oxidant concentration up to 12 mM. Beyond 12 mM, it was found to decrease. **Figure 2** shows the influence of H_2O_2 concentration in UV/H_2O_2, photo-Fenton and TiO_2 photocatalysis processes at an oxidant concentration of 12 mM. The decrement in % decolorization beyond 12 mM of oxidant was due to the scavenging action of using extra H_2O_2 or possible recombination of hydroxyl radicals. Sometimes, the excessive amount of H_2O_2 acts as a hole scavenger, which can produce per-hydroxyl radicals. The per-hydroxyl radicals are less reactive than hydroxyl radicals. Thus, it is always advisable to use an optimum dose of oxidant.

3.4 Pseudo-first-order reaction kinetics for hydroxyl and sulphate radical-based AOPs

The pseudo-first-order reaction kinetics was reported for the decolorization study of MGO dye in wastewater. Both processes, hydroxyl and sulphate radical-based AOPs, were employed to calculate the reaction rates and compared to study their efficiencies. The parameters of kinetic models for decolorization efficiency of MGO dye at 12 mM dose of H_2O_2 and 1.68 mM of persulphate concentrations using UV irradiation were studied with the application of linear regression analysis to [$\ln(C_0/C_t)$ versus t] data for pseudo-first-order model [16]. The obtained values of correlation coefficients for sulphate and hydroxyl radical-based AOPs were 0.9904 and 0.9261, respectively. The reaction rate constant values were 0.0276 min^{-1} for

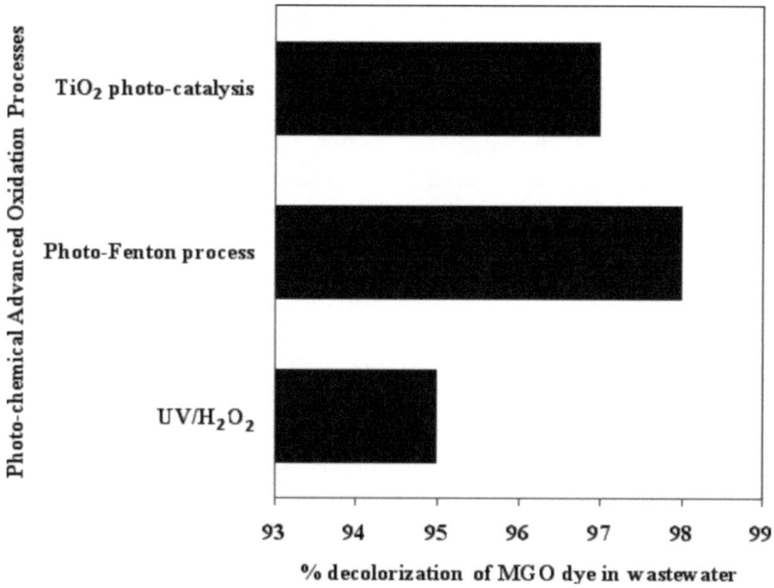

Figure 2.
Influence of H_2O_2 concentration in UV/H_2O_2, photo-Fenton and TiO_2 photocatalysis processes (experimental conditions: flow rate of 10 LPH, 12 mM of oxidant (H_2O_2), ferrous ion concentration of 60 mg/L, TiO_2 dosage of 0.6 g/L with irradiation of 32 W for 60 min of treatment time).

sulphate radical-based AOP and 0.0031 min^{-1} for hydroxyl radical-based AOP. This means that the decolorization was faster in sulphate radical-based AOP.

4. Utilization of different modified photocatalysts in photocatalysis

The use of different photocatalysts in the photocatalysis processes is a useful alternative and attractive option in advanced researches. The photocatalyst should possess higher photochemical stability and can effectually use UV irradiation for wastewater treatment. The word 'photocatalyst' is made up of two parts: photo and catalyst. The word 'photo' means light, and a 'catalyst' means the process in which the oxidation rate is enhanced with catalyst.

Activation energy 'Ea' was found to decrease with an increase in the rate of reaction. In addition, the photocatalysts possess some properties such as higher surface area, better crystallinity, low cost, non-toxicity, ease handling and cost-effective applicability [17]. Recently, the application of semiconductors like titanium dioxide (TiO_2) shows great potential in photocatalysis processes. This is because of its low cost and environmentally friendly nature. TiO_2 is generally used as a standard reference for the comparison of the photochemical activity of other catalysts. The fine powdered form of TiO_2 is normally used in the slurry form. This leads to the higher volumetric production of reactive oxygen species (ROS) that can be proportional to the number of active surface sites [18].

The post-separation of TiO_2 after the treatment is an additional step while using this photocatalyst in slurry form. It is an important step for avoiding any loss of catalyst and can be reused for the photooxidation of new pollutants. The catalyst can be recovered through conventional sedimentation or via crossflow filtration or some types of membrane filtrations. The applicability of immobilized form of photocatalysts can increase the operational difficulty as the photon might not penetrate to every single surface active site for its activation. Silver orthophosphate (Ag_3PO_4)

has also gained much attention for its use as photocatalysts as it shows higher photooxidative capabilities for O_2 evolution from the split-up of water and better removal efficiency for textile wastewater [19]. Approximately 90% of quantum efficiency can be achieved with Ag_3PO_4 at a wavelength of 420 nm in the oxidation of wastewater. During this phenomenon, $AgNO_3$ acts as a scavenger. This mentioned quantum efficiency was extremely higher than other photocatalysts, such as 20% in the case of N-doped TiO_2 and $BiVO_4$ photocatalysts. But the uncontrolled and unwanted photo-corrosion of Ag_3PO_4 is the main hurdle for its application in photocatalytic systems.

The photo-corrosion of Ag_3PO_4 happens due to its slighter solubility (0.02 g/L) in water and its characteristics of energy band structure. Thus, it's a big challenge for the researchers to improve and modify Ag_3PO_4 for their stability and to increase its activity without the help of any sacrificial reagents. Sometimes, the semiconductor, TiO_2, is doped to slow down the recombination of fast charge and to enable the visible light absorption with the generation of defect states in the bandgap. In the earlier case, the conduction band electrons or valence band holes are trapped in the defect sites, retarding their recombination and enhancing the interfacial charge transfer. In the latter case, the electronic transitions from the valence band to defect state or from defect state to conduction band are permitted under UV irradiation [20, 21].

The dopants are categorized into metal and nonmetal ions, and their selection is critical in considering their overall photocatalytic activity. Using noble metal ions (platinum, gold and silver) as dopants has gained less attention than the use of transition metals. Metallic doping serves the purpose of both a conduction band electron trapping and a cocatalyst in photocatalysis. Doping of TiO_2 is also possible with some nonmetallic elements such as carbon, sulfur, nitrogen and boron for studying their visible light photocatalysis.

The photocatalytic decomposition of water pollutants on nitrogen-doped TiO_2 under the illumination of visible light is essentially induced by the surface intermediates formed by oxidation of water or by the reductive activation of oxygen. It does not happen by the direct reactions with positively charged holes trapped at N_2-induced mid-gap level. Moreover, the use of the carbon-doped TiO_2 in photocatalysis is also controversial. Different methods have been proposed for the synthesis of C-doped TiO_2, such as carbon coating and carbon mounting. Two carbon dopants are there: the first one is anions that can substitute oxygen (O_2) in the lattice structure, and the second one is cations that can occupy an interstitial lattice site. The oxidation states that are linked with its bonding phases are +4 (carbonates having C-O bond) and −4 (carbides having Ti-C bond). These oxidation states are strongly dependent on the method of catalyst preparation and operating conditions used. Both states can be even copresent in the prepared photocatalysts. Sometimes, the TiO_2 is also doped with more than two elements for further enhancement in the photocatalytic activity under UV and visible light. This type of activity makes the doped TiO_2 more stable via compensation of charge. The metal/metal (Pt/Cr), nonmetal/nonmetal (N/S) and metal/nonmetal pair (Pt/N) combinations can be used further for doping [20]. Therefore, TiO_2 photocatalysts are doped with different elements, as discussed above, to overcome its demerits before use in oxidation, i.e. less absorbance with visible light and possible recombination of electron–hole pairs [22].

5. Conclusions

Advanced oxidation technologies represent a powerful option for the removal of organic pollutants in industrial wastewater. Different AOPs have been investigated, and thus, it allows the selection of the most suitable process for the specific wastewater treatments. Different photochemical AOPs also have much potential for the reduction of organics in wastewater. The combination of UV and radicals (hydroxyl or sulphate) can effectively eliminate the organics with higher removal efficiency than direct UV photolysis or persulphate oxidation alone. Decolorization of malachite green oxalate was performed with different homogeneous and heterogeneous AOPs. The influence of ferrous ions and oxidant concentration and comparative assessment of different photochemical AOPs are reported. 12 mM of H_2O_2 and 60 mg/L of ferrous ion concentrations have been observed to be optimum in the photo-Fenton process. In the comparative assessment studies of photochemical AOPs, the percentage dye decolorization efficiency has been decreased in the order: $UV/Fe^{2+}/H_2O_2$ > $UV/H_2O_2/TiO_2$ > UV/TiO_2 > UV/H_2O_2 > UV alone. The experimental results show that the MGO dye removal efficiency from wastewater is higher in the case of sulphate radical-based AOP than hydroxyl radical-based AOP. Although the pure form of TiO_2 is a well-liked photocatalyst for many reasons, it suffers from lower efficiency for photochemical oxidation and becomes deficient in the visible light activity that obstructs its practical applications. Thus, the surface-modified TiO_2 photocatalysts have been continuously investigated to overcome the shortcomings of pure TiO_2.

6. Future perspectives

The conceptual proof of advanced oxidation processes has already been offered after years of research for wastewater treatment. However, something is indeed missing, the application of AOPs from laboratory scale to large scale (i.e. totally operational use). The main hindrance is the level of operating cost (money required for per unit mass of organic pollutant eliminated or unit volume of wastewater treated) allied with AOPs in comparison with other traditional wastewater treatment techniques. How can advanced oxidation treatment techniques become more striking cost-wise? Some useful thoughts that can be considered for its broader applicability are as follows:

- The utilization of renewable energy is essential in this case. AOPs driven by sun irradiation have an apparent head start not only for the treatment of industrial wastewater but also for the production of energy simultaneously. This is the more economical solution for the issue of water purification and its reuse.

- Every time, AOPs should not be our first choice for the treatment of wastewater. It must be imposed to become selective, i.e. they must have a definite treatment target. Otherwise, valuable and expensive elements such as oxidants, energy and different catalysts are wasted without cause.

- Advanced oxidation technologies have been intensely benefitted from the advances in the field of material science. Novel materials have been identified continuously with better characteristics. Consequently, the use of AOPs for environmental applications is a topic in engineering, science and other multiple disciplines. All the researchers and scientists working on the same field from various disciplines must join forces to solve the problem successfully. Moreover, the use of waste materials as a catalyst, such as red mud (waste from bauxite processing industries), can reduce the cost of AOP treatment. This phenomenon is waste valorization. It is a new, impressive and somewhat unexploited concept.

So, it is not possible to replace accessible wastewater treatment technologies with AOPs from an economic standpoint. But this problem can be solved by using the combination of advanced oxidation processes with the ancient treatment technologies. It is feasible and cost-effective, also comparing to a fully AOP application.

Author details

Damodhar Ghime and Prabir Ghosh*
Department of Chemical Engineering, National Institute of Technology, Raipur, Chhattisgarh, India

*Address all correspondence to: prabirg.che@nitrr.ac.in

IntechOpen

© 2020 The Author(s). Licensee IntechOpen. Distributed under the terms of the Creative Commons Attribution - NonCommercial 4.0 License (https://creativecommons.org/licenses/by-nc/4.0/), which permits use, distribution and reproduction for non-commercial purposes, provided the original is properly cited. BY-NC

References

[1] Miklos DB, Remy C, Jekel M, Linden KG, Hubner U. Evaluation of advanced oxidation processes for water and wastewater treatment—A critical review. Water Research. 2018;**139**:118-131

[2] Dewil R, Mantzavinos D, Poulios I, Rodrigo MA. New perspectives for advanced oxidation processes. Journal of Environmental Management. 2017;**195**:93-99

[3] Jimenez S, Andreozzi M, Mico MM, Alvarez MG, Contreras S. Produced water treatment by advanced oxidation processes. Science of the Total Environment. 2019;**666**:12-21

[4] Contreras S, Pascual E, Esplugas S, Gim J, Rodri M. Comparison of different advanced oxidation processes for phenol degradation. Water Research. 2002;**36**:1034-1042

[5] Hassaan MA, El A, Madkour FF. Testing the advanced oxidation processes on the degradation of direct blue 86 dye in wastewater. Egyptian Journal of Aquatic Research. 2017;**43**:11-19

[6] Andreozzi R, Caprio V, Insola A, Marotta R. Advanced oxidation processes (AOP) for water purification and recovery. Catalysis Today. 1999;**53**:51-59

[7] Bedding ND, McIntyre AE, Perry R, Lester JN. Organic contaminants in the aquatic environment. Science of the Total Environment. 1982;**25**:143-167

[8] Luan M, Jing G, Piao Y, Liu D, Jin L. Treatment of refractory organic pollutants in industrial wastewater by wet air oxidation. Arabian Journal of Chemistry. 2017;**10**:S769-S776

[9] Amor C, Marchao L, Lucas MS, Peres JA. Application of advanced oxidation processes for the treatment of recalcitrant agro-industrial wastewater: A review. Water. 2019;**11**:1-31

[10] Wols BA, Hofman-caris CHM. Review of photochemical reaction constants of organic micropollutants required for UV advanced oxidation processes in water. Water Research. 2012;**46**:2815-2827

[11] Shu Z, Bolton JR, Belosevic M, Gamal M, Din E. Photodegradation of emerging micropollutants using the medium-pressure UV/H_2O_2 advanced oxidation process. Water Research. 2013;**47**:2881-2889

[12] Olmez-Hanci T, Arslan-Alaton I. Comparison of sulfate and hydroxyl radical based advanced oxidation of phenol. Chemical Engineering Journal. 2013;**224**:10-16

[13] Gu D, Guo C, Lv J, Hou S, Zhang Y, Feng Q, et al. Removal of methamphetamine by UV-activated persulfate: Kinetics and mechanisms. Journal of Photochemistry and Photobiology A: Chemistry. 2019;**379**:32-38

[14] Gao Y, Gao N, Chu W, Zhang Y, Zhang J, Yin D. UV-activated persulfate oxidation of sulfamethoxypyridazine: Kinetics, degradation pathways and impact on DBP formation during subsequent chlorination. Chemical Engineering Journal. 2019;**370**:706-715

[15] Huang J, Zhang H. Mn-based catalysts for sulfate radical-based advanced oxidation processes: A review. Environment International. 2019;**133**:1-23

[16] Ghime D, Ghosh P. Oxidative decolorization of a malachite green oxalate dye through the photochemical advanced oxidation processes. 2019;**21**:195-203

[17] Singh P, Shandilya P, Raizada P, Sudhaik A, Rahmani-sani A, Hosseini-bandegharaei A. Review on various strategies for enhancing photocatalytic activity of graphene based nanocomposites for water purification. Arabian Journal of Chemistry. 2018. DOI: 10.1016/j.arabjc.2018.12.001

[18] Nan M, Jin B, Chow CWK, Saint C. Recent developments in photocatalytic water treatment technology: A review. Water Research. 2010;**44**:2997-3027

[19] Chen X, Dai Y, Wang X. Methods and mechanism for improvement of photocatalytic activity and stability of Ag_3PO_4: A review. Journal of Alloys and Compounds. 2015;**649**:910-932

[20] Park H, Park Y, Kim W, Choi W. Surface modification of TiO_2 photocatalyst for environmental applications. Journal of Photochemistry and Photobiology C: Photochemistry Reviews. 2013;**15**:1-20

[21] Dedkova K, Matejova L, Matejova K, Peikertova P, Kutlakova KM, Kukutschova J. Study of the antibacterial activity of cerium doped TiO_2 photocatalysts. In: NANOCON 2013 Proceedings, Brno, Czech Republic, EU. Vol. 10. 2013. pp. 16-21

[22] Patil SB, Basavarajappa PS, Ganganagappa N, Jyothi MS, Raghu AV, Raghava K. Recent advances in non-metals-doped TiO_2 nanostructured photocatalysts for visible-light driven hydrogen production, CO_2 reduction and air purification. International Journal of Hydrogen Energy. 2019;**44**:13022-13039

Section 2

General Concepts of Heterogeneous Catalysis

Chapter 2

Heterogeneous Catalytic Process for Wastewater Treatment

Ting Zhang

Abstract

This chapter introduced heterogeneous catalysis and described diverse heterogeneous catalytic processes for wastewater treatment. The main advantages of heterogeneous catalysis were explained compared with homogeneous catalysis. The methods of synthesis and characterization of heterogeneous catalysts with some examples were then elaborated. The principle of heterogeneous catalytic treatment process of refractory wastewater was analyzed, and several different types of heterogeneous catalytic oxidation processes, technical progresses, and application examples were presented. The mechanisms of heterogeneous catalytic oxidation degradation of pollutants in wastewater were also discussed. According to the review, the heterogeneous catalytic oxidation technology was considered having a good application prospect, and the further research directions of heterogeneous catalysis were proposed.

Keywords: heterogeneous catalysis, synthesis, characterization, wastewater treatment, refractory wastewater, catalytic oxidation processes, mechanism

1. Introduction

Water pollution, especially organics pollution, has become a global environmental problem. The severity, nature, and harm to organisms and human beings of organic pollution are continually changing with the development of society and industry. In recent years, the pollution of refractory organics and their treatment have become a hot topic to researchers all over the world. Refractory organic pollutants, such as halogenated organics, surfactants, nitro compounds, heterocyclic compounds, phenolic compounds, and polycyclic aromatic hydrocarbon, among others, with high toxicity, are difficult to be degraded by microorganisms. Many of these pollutants have toxic effects on humans and organisms, such as carcinogenic, teratogenic, and mutagenic effects. The harmfulness of refractory organic compounds has posed a severe threat to human health and ecosystems. Therefore, how to control the pollution of refractory organics has always been an important research topic in the field of environmental protection.

Researchers have conducted in-depth studies on the treatment of wastewater containing refractory organics. Refractory organics can be removed by adsorption, advanced oxidation, and membrane separation, among others (**Figure 1**). Among these technologies, advanced oxidation processes (AOPs) have made remarkable progress and achievements in the treatment of wastewater containing refractory organics. AOPs, including photochemical oxidation, catalytic wet oxidation,

IntechOpen

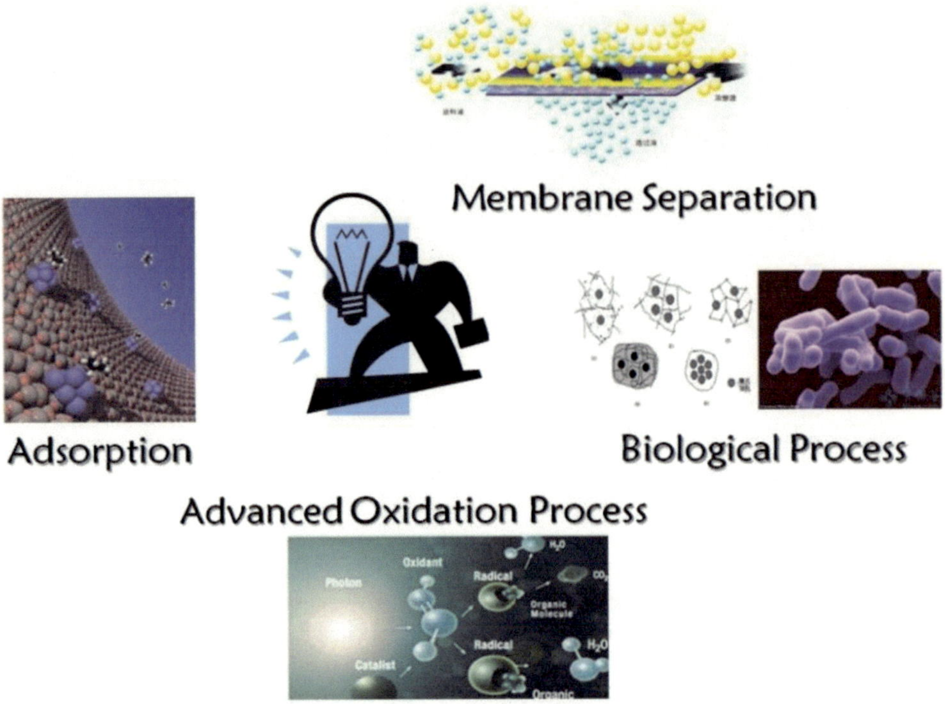

Figure 1.
Refractory organics removing methods.

sonochemical oxidation, electrochemical oxidation, and ozone oxidation, can decompose and transform toxic, harmful, and refractory macromolecular organics into nontoxic, harmless, and biodegradable small molecular organics. The final products of the oxidation are carbon dioxide, water, and inorganic ions, while no excess sludge is produced.

The catalysis term implies the process of increasing the rate of a chemical reaction by adding a catalyst, which is ultimately regenerated so that its amount remains unchanged. Catalysis is inhibited if the catalyst or the reactant is removed, or inhibitors alter the reaction. Inhibitors are materials that slow down the overall reaction by shortening the reaction chains, generally by reacting with one of the chemical components that maintain the chain and entering into a nonchain reaction. A wide variety of substances, such as phenols, sugars, and alcohols, have been found to act as inhibitors during the catalysis process.

The homogeneous catalysis occurs in a single phase, in which the catalyst is dispersed in an aqueous solution or gas mixture with the reactants. Conversely, heterogeneous catalysis occurs in more than one phase; for instance, the reactants are liquids, and the catalyst is solid [1].

AOPs' aim is to generate hydroxyl radicals ($^{\bullet}OH$), which can oxidize most chemical species, especially hard-to-degrade organics. Hydroxyl radicals are nonselective when oxidizing organics, while their oxidation potential has been estimated as 2.8 and 1.9 V at pH 0 and 14, respectively [2].

The heterogeneous Fenton reaction has demonstrated better catalytic performance and environmentally friendly characteristics when compared with the homogeneous Fenton reaction. Although homogeneous catalysts are generally very efficient for the degradation of organic compounds, the catalysts of iron ions are dissolved in water, their separation and reuse are rather difficult at the end of treatment, and ferric ions are generated as by-products. The removal of iron ions from the treated water needs a large number of chemicals and manpower, which will increase the cost of treatment. Another drawback is the tight range of pH in reactions of homogeneous Fenton systems. The pH of the solution in a Fenton system should be adjusted to around 3.0 before carrying out the wastewater treatment, and the reagents for acidification are very costly [3].

In this chapter, the preparation processes and characterization methods of heterogeneous catalysts, as well as treatment applications, are reviewed. This chapter aims to provide a basis for further studies on new heterogeneous catalysts for the treatment of refractory organics.

2. Heterogeneous catalysis

Many catalytic processes, in which the catalyst and the reactants are not present in the same phase, are heterogeneous catalytic reactions. They include reactions of gases (or liquids) at the surface of a solid catalyst. The surface is where reactions take place, so the catalysts are generally prepared to provide large surface areas per unit. Metals, metals coated onto supporting materials, metalloids, metallic films, and doped metals have all been used as heterogeneous catalysts.

With solid catalysts, at least two of the reactants are chemisorbed by the catalyst, and they will react at the surface of the catalyst, with which the products are formed as readily as possible. Then, the products are released from the catalyst surface, as can be seen in **Figure 2**.

Heterogeneous catalysts can be divided into two categories: unsupported catalysts (bulk catalyst) and supported catalysts. Unsupported catalysts include metal oxide catalysts, such as Al_2O_3, SiO_2, and B_2O_3, and zeolite molecular sieve catalysts, such as ZSM-5, X, Y, and B types of zeolite. Supported catalysts often employ porous materials as supporters, and active components, metals or metal oxides, for example, are coated on the surface of the supporters to form the catalysts. Heterogeneous catalysts can accelerate many chemical processes, including CO oxidation, selective oxidation, water-gas shift, selective hydrogenation, electrocatalysis, photocatalysis, organic reactions, $deNO_x$, and reforming reactions, among others, as shown in **Figure 3**.

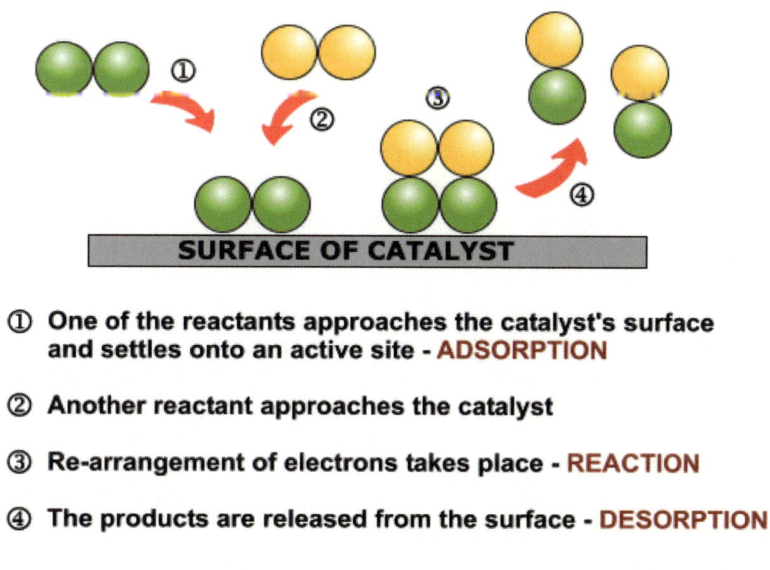

Figure 2.
Catalytic processes on a solid catalyst.

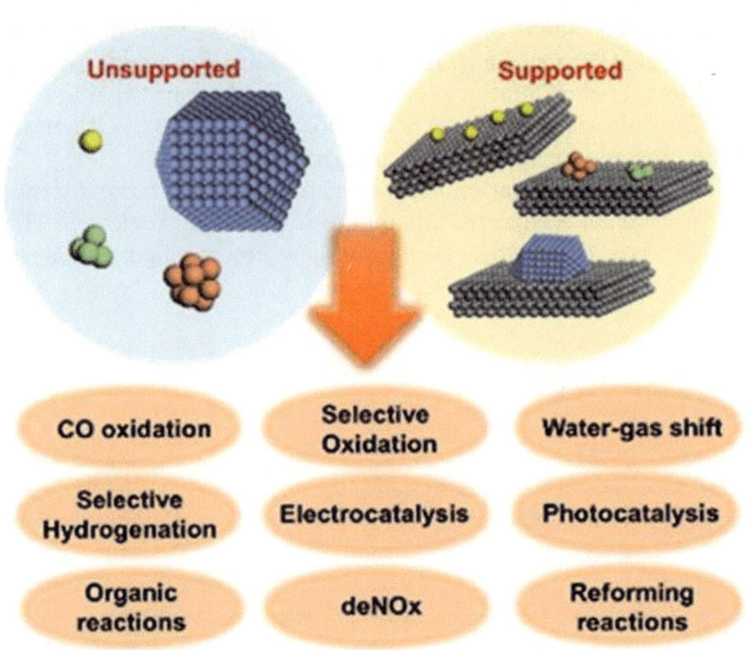

Figure 3.
Types of heterogeneous catalysts and their usages.

The heterogeneous catalytic process, as an efficient green method coping with organic wastewater, has attracted considerable attention in the last two decades. The highly reactive and nonselective hydroxyl radicals can oxidize and mineralize most organic compounds at near diffusion-limited rates, mainly unsaturated organic compounds.

3. Synthesis and characterization of heterogeneous catalysts

Catalysis is a surface phenomenon occurring on the surface or interface; in order for the catalyst to have more active sites, it is necessary to maximize its surface area, that is, to increase the porosity of solid catalysts. There are two ways to increase porosity. First, if the solid catalyst is composed of nonporous crystals, the catalyst should be prepared with the smallest grain size (i.e., to increase its external surface area). Second, if the solid catalyst is composed of grains with regular channels, the catalyst should be prepared by particles with regular channels (i.e., to increase its internal surface area). **Figure 4** shows the schematic of various catalyst development strategies, which aim to increase the number of active sites or increase the intrinsic activity of each active site.

To characterize heterogeneous catalysts, BET surface area, scanning electron microscope (SEM), transmission electron microscope (TEM), Fourier transform infrared spectroscopy (FT-IR), and X-ray diffraction (XRD) are commonly used characterizing methods.

The specific surface areas and pore volumes of the samples are often determined by physisorption of nitrogen at −196°C using a Micromeritics ASAP instrument. The specific surface area is calculated using the BET method and pore size distribution was obtained by using the BJH method.

A field-emission scanning electron microscope is utilized to determine the crystal morphology and chemical element composition. Measurements are made with

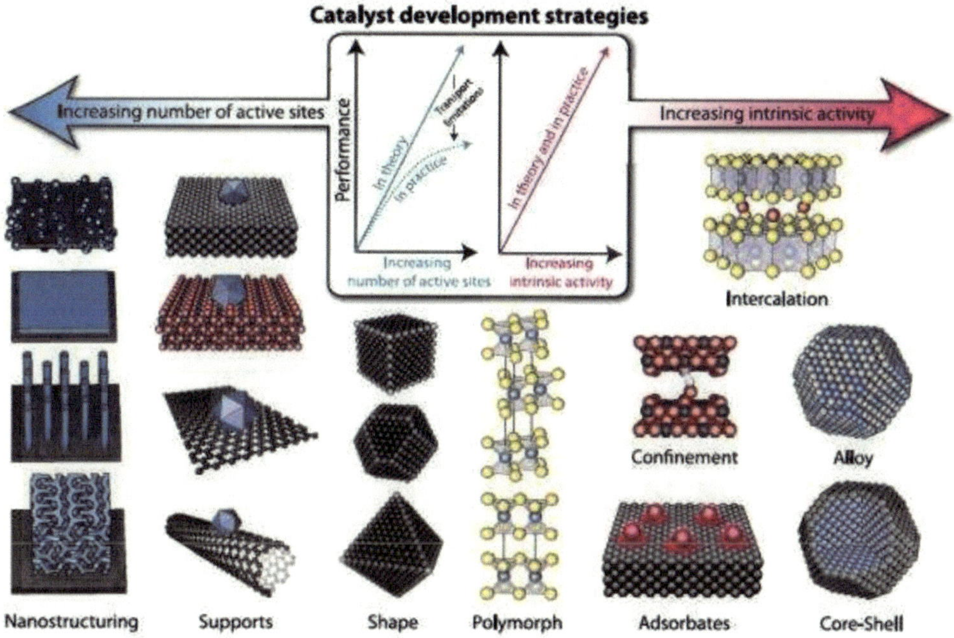

Figure 4.
Catalyst development strategies.

SEM using a digital imaging processing at room temperature and acceleration voltage of 20 kV. The samples are usually prepared as electrically conductive by sputter coating with a thin layer of gold under vacuum conditions to avoid charge accumulations of nonmetal catalysts, and all samples are dispersed in ethanol before tests.

A transmission electron microscopy is utilized to observe and analyze the micromorphology and microstructure. FT-IR spectroscopy is used to confirm the chemical structure, any changes in the compositional or functional group during the preparation of catalysts. Measurements are made on a FT-IR spectrometer after samples are mixed with 300 mg of spectroscopic grade KBr and ground in an agate mortaring in the range of 4000–400 cm^{-1} at room temperature.

To investigate the crystalline structure and stability of catalyst composites, powder X-ray diffraction patterns are recorded by X-ray diffractometer using Cu K_α radiation (λ = 0.1542 nm) at a rate of 0.02°/s in the range of 5–80° with an operating voltage of 40 kV and electric current of 150 mA.

If the catalysts contain metals, XPS (X-ray photoelectron spectroscopy) of the samples are often tested. XPS was used to identify metal oxidation states of the composites.

3.1 Bulk catalysts

Bulk catalyst refers to the whole catalyst particle, its external surface, and internal components, where almost all of them are active substances. Some examples of bulk catalysts are silica-aluminum catalysts for catalytic cracking and ammonia synthesis, zinc-chromium oxide or copper-zinc-aluminum catalysts for hydrogen production, and iron molybdate catalysts for methanol oxidation [4–7]. The main chemical processes for the preparation of such catalysts are precipitation, gelation, and crystallization. **Figure 5** shows the synthesis processes of bulk catalysts.

$LiFe(WO_4)_2$ catalyst for decolorization of methylene blue is an example of bulk catalysts. $LiFe(WO_4)_2$ particles are prepared using solid-state reactions. The

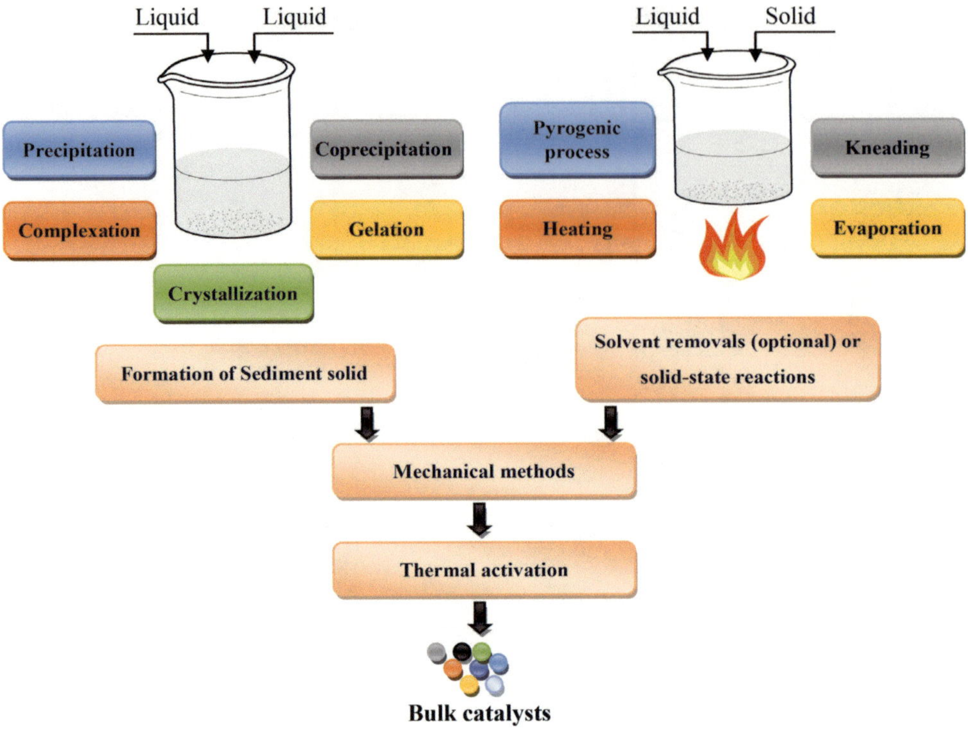

Figure 5.
Synthesis process of bulk catalysts.

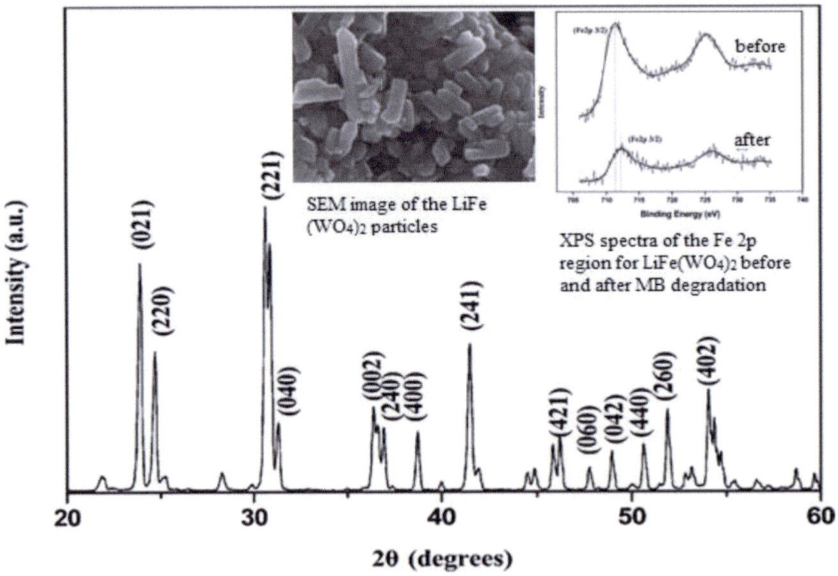

Figure 6.
Characterizations of LiFe$(WO_4)_2$ catalyst [8].

powders, consisting of Li_2CO_3, Fe_2O_3, and WO_3 with a molar ratio of 1:1:4, are mixed uniformly in a glass mortar. Then, the mixture is mechanically milled with hydrous ethanol as a lubricating agent. Afterwards, the powders are dried to remove the ethanol. Finally, the powder is calcined and yields well-crystallized LiFe $(WO_4)_2$ [8].

The characterizations of $LiFe(WO_4)_2$ can be seen in **Figure 6** [8]. The XRD patterns of the $LiFe(WO_4)_2$ show that the peaks of the sample match well with the published standard data (PDF 01-072-0751), and $LiFe(WO_4)_2$ has a highly crystalline and single-phase structure, without any peaks of unreacted precursors. The morphological characteristic of the $LiFe(WO_4)_2$ particles shows that the LiFe $(WO_4)_2$ crystal was rod-like, that the average length of the samples varied over the range of 300–500 nm, and that the width of the sample was about 200 nm. XPS spectra show that the binding energy of 711.94 eV belongs to Fe2p3/2. The binding energy value slightly decreased after 1 h reaction of MB decolorization due to the transformation of Fe(III) to Fe(II) after heterogeneous photo-Fenton reaction [8].

3.2 Supported catalysts

Supported catalysts are the most often used catalysts in wastewater treatment processes. They include the surface-loaded catalysts, core-shell catalysts, and thin-film catalysts.

3.2.1 The active catalytic components loaded onto the carrier surface

Surface-loaded catalysts mean active catalytic components loaded onto the carrier, which can be divided into two categories: one is concentrating the active components to the surface of porous solid carriers by means of dipping the carriers into the precursor contained ions or molecules and the other is that the predecessors or catalyst solid particles are adhered or deposited to the surface of the carrier, such as sol–gel method.

Carriers are inorganic porous materials, such as alumina, silicon oxide, activated carbon, zeolite, molecular sieve, and clays, among others. Commonly used catalyst carriers' structure data are shown in **Table 1**.

The most important supported catalysts are metal- or oxide-supported catalysts. Metal precursors are usually metal salts, such as nitrates, carbonates, sulfates, and chlorides. These salts are also the raw materials for preparing metal oxide supporters and bulk catalysts. When they dissolve in water, they will be dissociated into ions, and metal ions will further form hydrated ions. If the catalyst supporters are added to these metal salt solutions, the active components can be loaded onto the supporters by impregnation or sol–gel methods, thereby obtaining heterogeneous catalysts. **Figures** 7 and **8** are typical processes of impregnation and sol–gel methods for heterogeneous catalyst synthesis.

Carriers		Specific surface area/(m^2/g)	Pore volume/(mL/g)
Oxide	Al_2O_3	200~500	0.3~0.9
	SiO_2	400~800	0.4~4.0
	SiO_2-Al_2O_3	400~600	0.5~0.9
	MgO	30~140	0.3
Carbon	Activated carbon	~1000	0.3~2.0
Clay	Bentonite	150~280	0.3~0.5
	Attapulgite	100~200	0.3~0.5

Table 1.
Structure data of commonly used catalyst carriers.

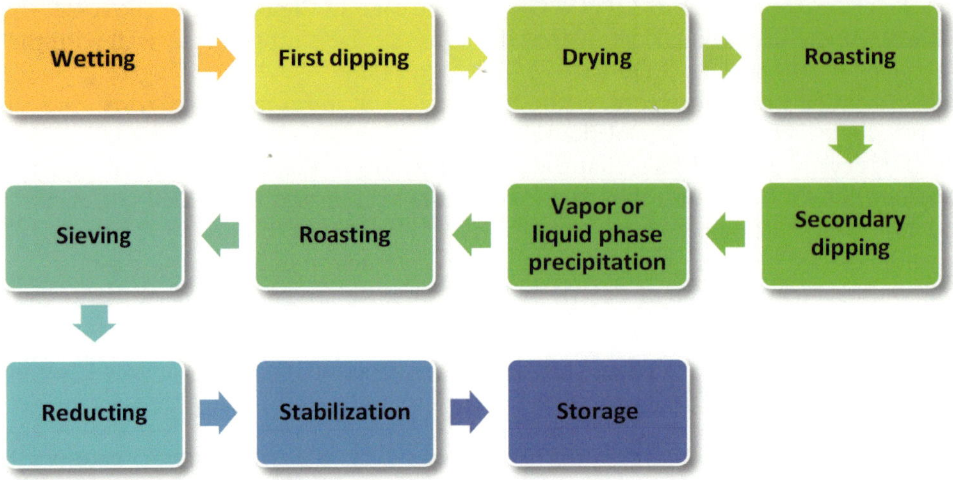

Figure 7.
Typical process of impregnation for heterogeneous catalyst synthesis.

3.2.1.1 Carbon-based catalysts

Carbon is often used as a catalyst supporter due to its large surface area and porous structure. There are many types of carbon: activated carbon, activated carbon fiber, mesoporous carbon, and multiwalled carbon nanotubes, among others. Here are some examples of carbon supporters for heterogeneous catalysts.

Ordered mesoporous carbon-supported iron catalysts (Fe/OMC) were prepared by the incipient wetness impregnation method [9]. OMC was prepared by triconstituent coassembly of resols, oligomer silicates from TEOS, and triblock copolymer F127 template, followed by carbonization and silica removal. The OMC was impregnated in an iron nitrate solution; then, the catalyst was dried, followed by calcination under N_2 flow.

Characterizations of Fe/OMC are shown in **Figure 9** [9]. From the XRD patterns, all of the samples show a broad peak at 23.5° corresponding to amorphous carbon. The crystallinity of iron oxides is improved with the increase of calcination temperature. TEM images showed a highly ordered array in the structure of OMC and Fe/OMC catalysts, which suggests the preservation of the long ordered arrangement of the channels after the iron deposition. The iron oxide particles are not clearly seen in the TEM image at a low calcination temperature of the samples, as the calcination temperature increases, the iron particles become more prominent and are revealed in the TEM images. The iron particles are well dispersed from the TEM pictures [9].

Fe_3O_4/multiwalled carbon nanotubes (MWCNTs) were synthesized by in situ chemical oxidation coprecipitation. MWCNTs were pretreated by stirring in a flask, containing a mixture of sulfuric acid/nitric acid and refluxing in an ultrasound wave cleaner. The as-treated MWCNTs were suspended in deionized water and put into a 95°C water bath. $FeSO_4{\cdot}7H_2O$ was added into the MWCNT suspended solution and NaOH and $NaNO_3$ solution was added dropwise, which is heated up to 95°C into the heating MWCNT-metal solution with vigorous stirring as well as stable N_2 flow during the entire reaction period. After 2 hours, the Fe_3O_4/MWCNT nanocomposites were formed [10].

Characterizations of Fe_3O_4/MWCNTs are shown in **Figure 10** [10]. The TEM images show the octahedron Fe_3O_4 nanoparticles with diameters ranging from 40 to 100 nm growing on the MWCNTs surface regularly and most of the Fe_3O_4 nanoparticles were strung by MWCNTs. The room-temperature Raman spectra

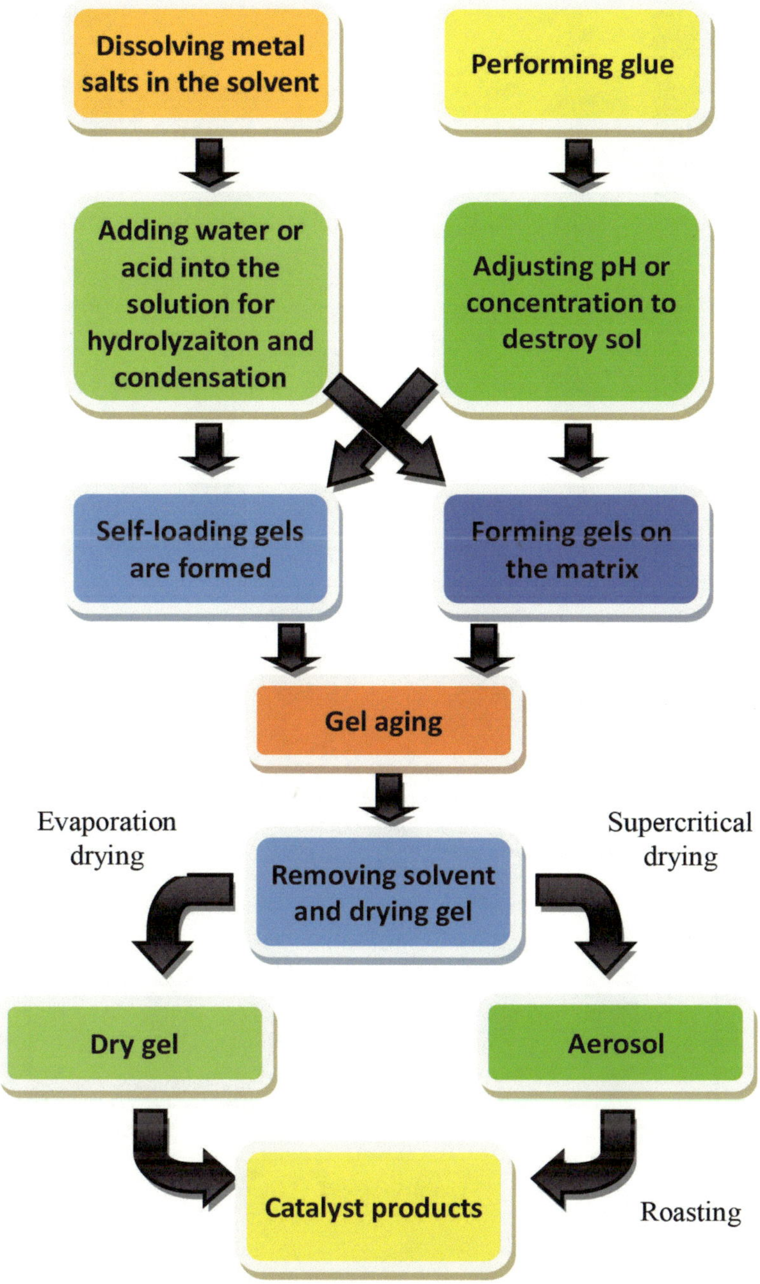

Figure 8.
Typical process of the sol–gel method for heterogeneous catalyst synthesis.

show a difference in the crystallinity of MWCNTs before and after the Fe_3O_4 loading process. Some extra peaks were observed at lower wave numbers corresponding to vibration modes of Fe–O bonds of Fe_3O_4 nanoparticles, and Fe–C bonds confirm the formation of Fe_3O_4 nanoparticles on the surface of MWNTs [10].

3.2.1.2 Oxide-based catalysts

Recently, metal-containing oxide-based (mesoporous silica, Al_2O_3, and ZnO, among others) catalysts have attracted much attention because the oxide-based

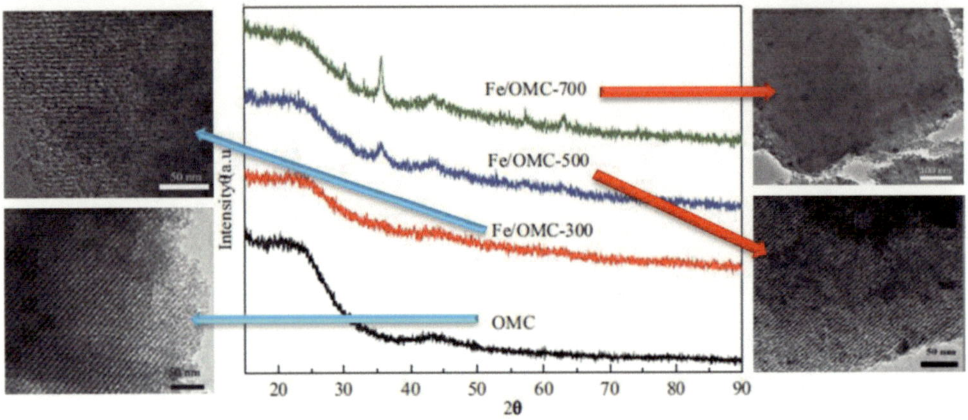

Figure 9.
Characterizations of Fe/OMC [9].

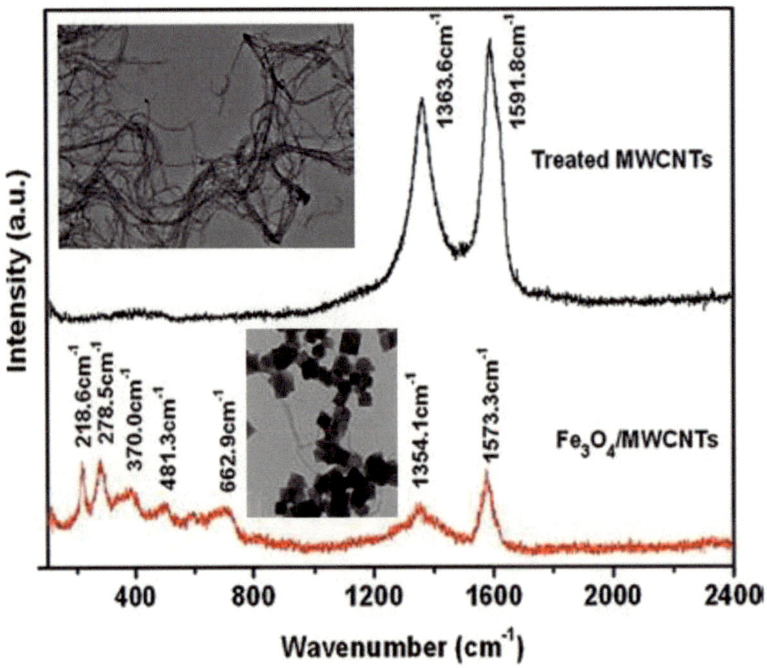

Figure 10.
Characterizations of Fe_3O_4/MWCNTs [10].

materials have a high surface area, good thermal stability, and favorable hydrothermal stability. Numerous oxide-based composites were synthesized for degrading organic compounds such as methyl orange, acid orange, polyarylamide, and phenol [11–16].

The porous $Fe_2O_3{\cdot}SiO_2$ composite was prepared by a facile impregnation method [17]. Typically, silica powder was immersed in $Fe(NO_3)_3$ and the mixture was vigorously stirred for 24 h at room temperature; then, the sample was dried and calcined. The obtained powder was cooled to room temperature to get as-synthesized $Fe_2O_3{\cdot}SiO_2$ for catalytic experiments.

Characterizations of $Fe_2O_3{\cdot}SiO_2$ are shown in paper [17]; those figures present the FE-SEM and TEM images, EDS spectrum and EDS mapping image, XRD patterns, N_2 adsorption/desorption isotherms, and FT-IR spectra of as-synthesized $Fe_2O_3{\cdot}SiO_2$ and as-synthesized SiO_2. As shown in SEM images, silica is like light clouds and showed an aggregate shape, while the as-synthesized $Fe_2O_3{\cdot}SiO_2$ seems to become darker and denser after iron-loading, and $Fe_2O_3{\cdot}SiO_2$ composite exhibited higher aggregation than silica due to impregnation and calcination processes, as can be seen in TEM images. Both samples were a bulk shape, which consisted of many small particles, approximately 10–20 nm in size. In TEM images, the iron oxide nanoparticles with about 5 nm in size were well dispersed on silica particles, while the iron oxide was not clearly distinguished from the silica in SEM images. The elemental composition of $Fe_2O_3{\cdot}SiO_2$ composite from EDS stated that the composite contained 8.1 wt.% Fe and showed the dispersion of Fe on the surface of SiO_2. The wide-angle XRD diffractions of SiO_2 and $Fe_2O_3{\cdot}SiO_2$ composite showed that the diffraction peak of 20.1° belongs to amorphous silica, and the diffraction peaks at 21.8, 25.0, 35.39, and 49.3° belong to α-Fe_2O_3. The broad and low-intensity peaks of Fe_2O_3 could result from low loading deposition level, well dispersion, and poor crystalline structure with a small size of metal oxide particles in the composite. According to the N_2 adsorption/desorption isotherms of as-synthesized SiO_2 and as-synthesized $Fe_2O_3{\cdot}SiO_2$ and IUPAC classification, both samples clearly showed a type II with an H1 hysteresis loop indicating material with agglomerates or compacts of approximately uniform spheres. The FT-IR spectra of the sample showed the small bands found at 685 and 604 cm^{-1} could be assigned to stretching vibrations of Fe-O and Si-O-Fe, respectively, which confirmed that SiO_2 indeed binds with Fe_2O_3 [17].

3.2.1.3 Clay-based catalysts

Clays such as montmorillonite, kaolin, and bentonite [18–20] are often employed as supporters in the use of heterogeneous catalysts, in order to lower the treatment cost of wastewater. Among the perspective clays, attapulgite clay is also a promising choice. Attapulgite clay is a kind of clay mineral with attapulgite as its main component. It is a crystalline hydrated magnesium aluminum silicate mineral with a unique layer-chain structure. Due to its particular structure, attapulgite has various excellent properties, such as good adsorption and catalytic properties [21–23]. Some examples of ATP for catalyst supporters are as below.

The first example is the preparation of the Fe_3O_4/ATP catalyst [24]. ATP clay was mixed with water evenly and aged 24 h at room temperature. The aged ATP was granulated to microparticle and dried and then roasted. Spherical ATP particles were put into the mixture solution of thoroughly mixed oleylamine (or ethanol), and n-hexane was added with $Fe(acac)_3$ and immersed for a particular time in an ultrasonic instrument. After separating ATP particles from the mixture solution, they were dried at a vacuum drying oven and then the particles were put in a muffle furnace and roasted without air. Thus, the particle heterogeneous catalyst Fe_3O_4/ATP was obtained. The process of preparing the Fe_3O_4/ATP catalyst can be seen in **Figure 11**.

The characterizations of Fe_3O_4/ATP are shown in **Figure 12**. Rod-shaped particles with lengths of 500∼700 nm and widths of 100∼150 nm are visible in SEM micrographs. After the introduction of iron species into ATP, nano-Fe_3O_4 particles were coated evenly onto the surface of ATP.TEM micrographs confirmed that the Fe_3O_4 nanoparticles with an average diameter of 10∼25 nm stuck to the attapulgite rod-like fibers. The FT-IR results of Fe_3O_4/ATP show some peaks at lower wave numbers (480 and 565 cm^{-1}) corresponding to vibration modes of Fe-O bonds of Fe_3O_4 nanoparticles on the surface of ATP. XRD patterns of Fe_3O_4/ATP are contrasted with those of ATP. Sample Fe_3O_4/ATP showed the characteristic peaks

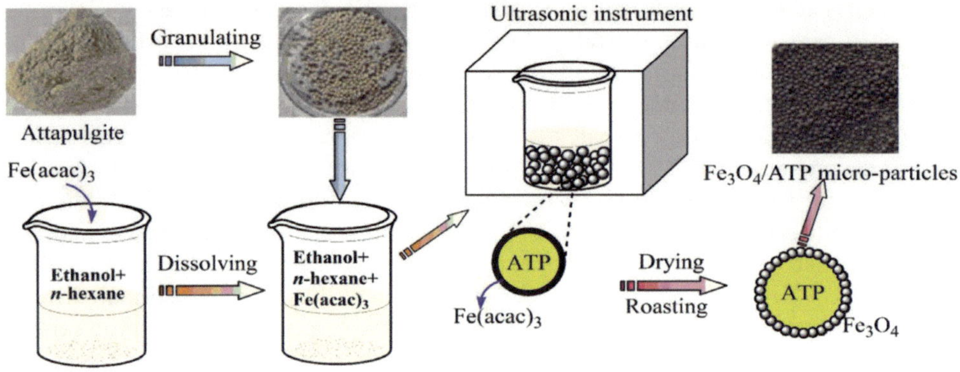

Figure 11.
Schematic of preparing Fe_3O_4/ATP catalyst [24].

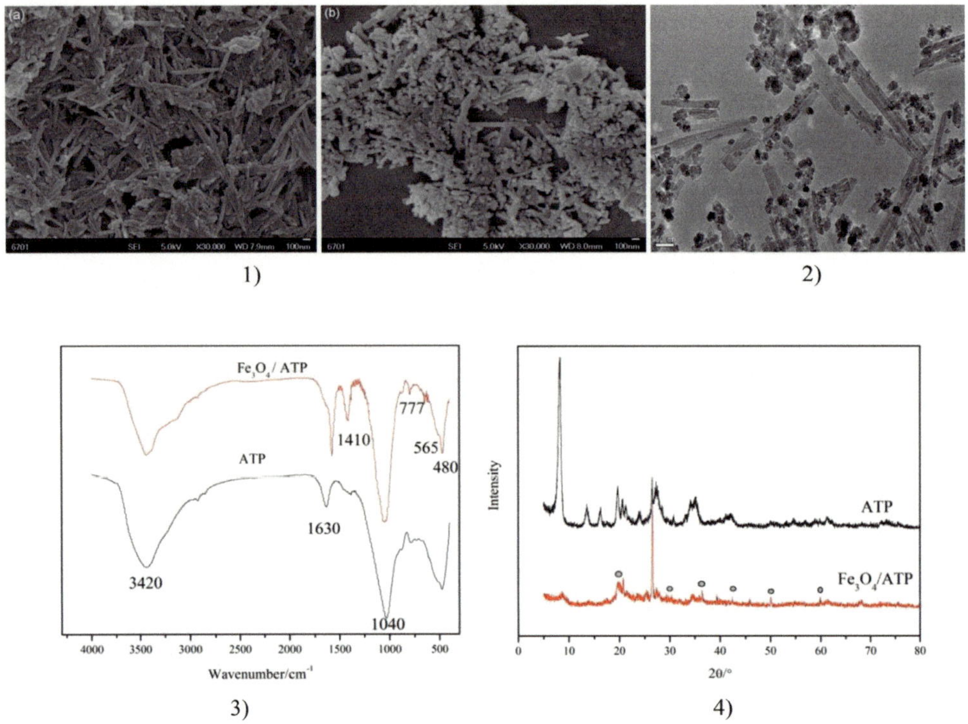

Figure 12.
Characterizations of Fe_3O_4/ATP [24]. (1) SEM micrographs of samples (a) ATP, (b) Fe_3O_4/ATP. (2) TEM micrographs of Fe_3O_4/ATP. (3) FT-IR spectra of samples. (4) XRD patterns of samples.

(5.3, 8.4, 19.7, 27.5, 34.6, and 42.6°) of cubic spinel structure known from bulk Fe_3O_4 phase [24].

Some researchers organo-modified attapulgite to make it have a larger surface area and more pores. Fe/OATP is a kind of catalyst for HA-Na degradation, which employed organo-modified attapulgite as a supporter [25]. Attapulgite (ATP) was firstly acidified with 1 mol/L of HCl solution, then washed with distilled water three times, and dried. Octadecyl trimethyl ammonium chloride (OTAC) was weighed according to a particular proportion and dissolved in distilled water. Then acidified ATP was added into the OTAC solution at 60°C. The mixture was stirred and treated in an ultrasonic instrument for 15 minutes. After that, the clay was washed with distilled water three times and dried in vacuum to obtain the organo-modified attapulgite (OATP). OATP and $FeSO_4$ were put into a beaker containing distilled water and stirred continuously. Sodium borohydride ($NaBH_4$) was slowly put into the beaker and stirred continuously. A large number of black floccules were produced. They were filtered and washed with ethanol for at least three times and dried in a vacuum drying oven to obtain the catalyst Fe/OATP. The synthesis process of Fe/OATP can be seen in **Figure 13**.

Characterizations of Fe/OATP are shown in **Figure 14** [25]. From SEM images, we can see that the length of the rod crystal had not changed after modification, but the layered structure, the surface area, and pore volume of the particles increase significantly, which means that the organic modification successfully enlarges the specific surface area of the ATP particles. After introducing nano-Fe into modified ATP particles (OATP), the spherical material with a diameter of about 80 nm is loaded on the rod structure of OATP, which showed that the nanoiron was loaded on the structure of OATP successfully. Some peaks, observed at 2923 and 2850 cm^{-1} in FT-IR curves of OATP and Fe/OATP samples, are symmetric stretching vibration peaks of $-CH_3$ and $-CH_2$, respectively. These peaks indicate that organics successfully modify attapulgite clay. After introducing the Fe species, the peaks at 2923 and 2850 cm^{-1} become weak. The strong peaks at 5.3, 19.7, 34.6, 35.2, 42.6, and 78.7° could be found in the XRD patterns of all these three samples, which means the crystal structure of ATP has not been broken during the process of modification or preparation. The characteristic peaks at 50.4, 61.9, and 73.14° shown in the sample of Fe/OATP indicate that Fe^0 was successfully located on the clay structure [25].

3.2.2 Core-shell structured catalysts

In recent years, the preparation and application of core-shell structured materials have attracted extensive attention [26–28]. The catalysts compounded with the core-shell structure often exhibit high overall activity, large specific surface area, excellent shape-selective catalytic effect, and good thermal stability in catalytic reactions [29, 30]. The existence of shell layers not only provides available active

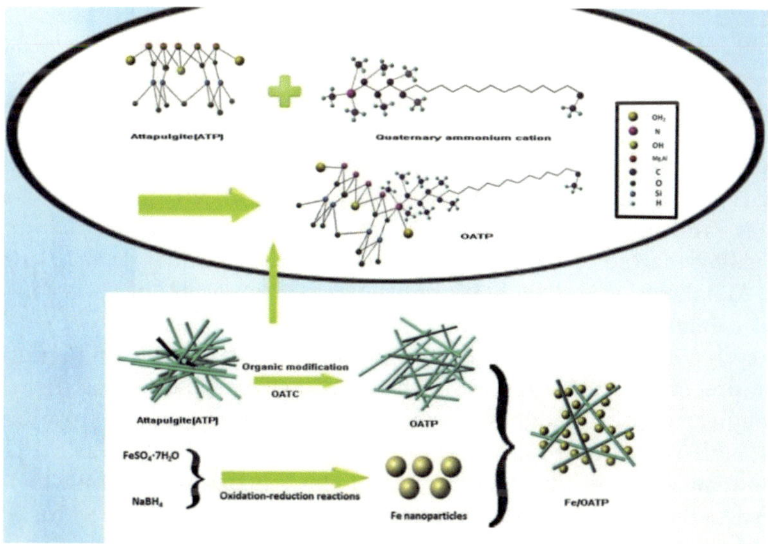

Figure 13.
Synthesis process of Fe/OATP [25].

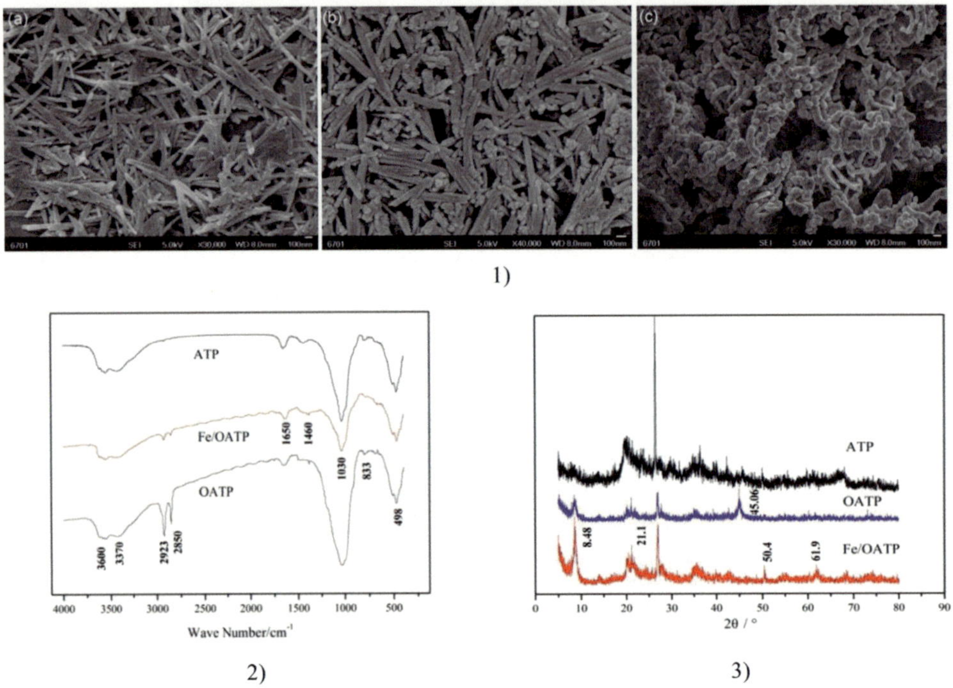

Figure 14.
Characterizations of Fe/OATP [25] (1) SEM images of the samples (a) ATP (b) OATP (c) Fe/OATP. (2) FT-IR spectroscopies of the samples (ATP, OATP, and Fe/OATP). (3) XRD patterns of the samples (ATP, OATP, and Fe/OATP).

sites for catalytic reactions but also protects for the core to avoid the collapse or sintering of the core catalyst structure leading to deactivation [31, 32].

Moreover, by selecting the chemical compositions of the core and shell, several inorganic substances with different characteristics can be incorporated into core-shell structure catalysts by appropriate methods, so that it can reach the goals of dual-function catalysis or even multifunction catalysis that single-component catalysts cannot achieve. It dramatically simplifies the experimental steps and saves precious time for researchers. Therefore, core-shell catalysts play an increasingly important role in the field of catalysis [33, 34]. The core-shell catalysts can comprise solid cores (such as inorganic, metal, or metal oxides) coated with polymers, inorganics, metal, or metal oxides [35–37]. The shell material can change the functionality, reactivity, and charge of the core surface, and also can strengthen the dispersibility and stability of the colloidal core. Catalytic or magnetic functions may be easily imparted to the colloidal core matter, depending on the coating materials. Metal oxides and zeolites are often used supporters of core-shell structured catalysts.

$Ag@Fe_2O_3$ core-shell structured catalysts were prepared using a homogeneous deposition precipitation method [38]. The core of silver colloids was first synthesized through a chemical reduction method using reagent grade silver nitrate as a precursor, $NaBH_4$, as the reducing agent and polyvinylpyrrolidone as the protecting agent. Then, an aqueous solution of urea and variable quantities of $Fe_2(SO_4)_37H_2O$ were added to the solution of silver colloids. The mixture was heated for 1 h and then cooled to room temperature. The precipitate was washed thoroughly with warm water to remove as much sulfate ions as possible and then separated by centrifugation. At last, the product was dried. As shown in **Figure 15**, silver colloids

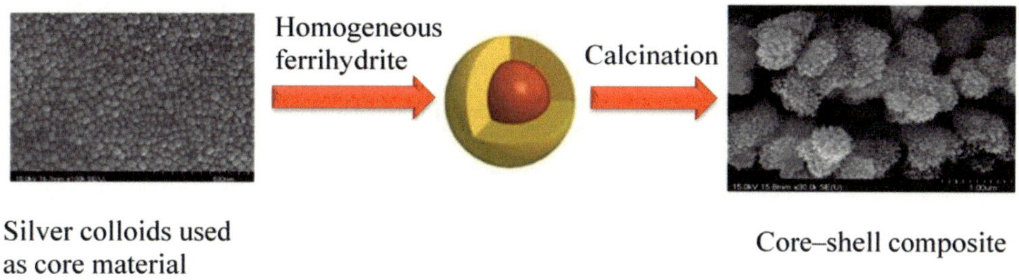

Figure 15.
Formation of Ag@Fe$_2$O$_3$ core-shell composite [38].

with about 40–50 nm in diameter were used as cores and the homogeneous ferrihydrite was used as shell. After being calcined, the rice-shaped Ag@Fe_2O_3 was formed, and the diameter and length of the Ag@Fe_2O_3 core-shell composite were about 10 and 20 nm, respectively. The calcined core-shell composites were aggregated with the size of 300–500 nm, which showed the porous nature of the Ag@Fe_2O_3 core-shell composite for catalytic reactions [38].

3.2.3 Thin-film catalysts

Thin films are a reliable and potential way of structuring the materials with tunable properties and flexible device construction in the field of energy and environment. It has gained technological importance as it helps to miniaturize the devices. Thin films are mostly modification methods to change the physical structure of materials. For example, the desired photocatalytic materials can be grown into thin films through the process of depositing molecular phase materials onto a substrate to form a solid material. The deposited thin films also can be grown along with the substrate and peeled off from the substrate and utilized as free-standing thin films. Thin-film catalysts have the potential to be used in the industries because synthesis processes of thin-film catalysts are simple in the handling of catalytic materials in several aspects such as their convenient design, adaptable form for the reactors, recyclability, recovery, etc. Therefore, the design of catalytic thin-film catalysts would be ideal for their wide applications.

Thin-film catalysts with good quality could be obtained by two methods: chemical depositions and physical depositions. Thin-film catalysts may consist of a layer and the substrate where the films are deposited on it, or consist of multiple layers. The following table shows different methods of thin-film deposition (**Table 2**) [40].

Thin-film catalysts can be used for hydrogen generation, photocatalytic CO_2 reduction to valuable chemicals, water treatment of toxic organic dyes, and bacterial disinfection. **Table 3** shows some examples of thin-film catalysts for water treatment.

For instance, Islam et al. have prepared zinc oxide (ZnO) and aluminum-doped zinc oxide (AZO) thin films onto a microscopic glass substrate using the sol–gel dip-coating technique at room temperature [49]. For both doped and undoped ZnO thin films, precursor solutions were prepared by dissolving [$Zn(CH_3COO)_2{\cdot}2H_2O$] in ethanol followed by stirring for 1 h; then, DEA was added as a stabilizer and water was added for hydrolysis. The resultant solution was stirred at room temperature for two more hours to obtain a clear and homogeneous sol. For AZO film, sols with different concentration of Al was prepared by dissolving an appropriate amount of [$AlCl_3{\cdot}6H_2O$] and [$Zn(CH_3COO)_2{\cdot}2H_2O$] in ethanol. In both cases, the precursor solution was deposited on clean glass substrates by the dip-coating technique. The dip-coated thin films were dried at room temperature

Physical deposition	Evaporation techniques	Vacuum thermal evaporation
		Electron beam evaporation
		Laser beam evaporation
		Arc evaporation
	Sputtering techniques	Direct current sputtering
		Radiofrequency sputtering
Chemical deposition	Sol–gel technique	
	Chemical bath deposition	
	Spray pyrolysis technique	
	Plating	Electroplating technique
		Electroless deposition
	Chemical vapor deposition (CVD)	Low pressure chemical vapor deposition (LPCVD)
		Plasma-enhanced chemical vapor deposition (PECVD)
		Atomic layer deposition (ALD)

Table 2.
Different methods of thin-film deposition [39].

Thin films	Applications	Reference
TiO_2-anatase films	Trichloroethylene degradation	[41]
Au-buffered TiO_2 thin films	Methylene blue degradation	[42, 43]
TiO_2/SiO_x	Dye degradation	[44]
ZnO thin films	Methylene blue degradation	
ZnO deposited on silicon	Dye degradation	[45]
ZnO nanorods grown on ZnO film	Dye degradation	[45]
α-Fe_2O_3 thin films on Si(100) substrates	Dye degradation	[46]
$SrTiO_3\alpha$-Fe_2O_3 thin films	Dye degradation	[47]
PbO/TiO_2	Stearic acid degradation	[48]

Table 3.
Different thin-film catalysts with application in wastewater treatment.

followed by calcinations at 400°C for 1 h to remove the organic compounds from the thin films.

Characterizations of Al-doped ZnO thin film can be seen in **Figure 16(a)** [49]. The crystal structure and different structural parameters of the films were analyzed from the X-ray diffraction (XRD) spectrum of the ZnO and AZO thin films. The undoped ZnO exhibits well-defined diffraction reflections corresponding to the wurtzite hexagonal phase of ZnO lattice planes. The intensity of all peaks decreases with Al doping suggesting deterioration of the crystallinity of ZnO films due to the incorporation of defects in the lattice site. From FE-SEM images, it can be seen that nanoparticles are formed and deposited on the film depositing area in all cases. For the pure ZnO thin film, the nanoparticle size varies from 10 to 70 nm. After the incorporation of Al, the particle size becomes more uniform, and the distribution of size is between 15 and 30 nm [49].

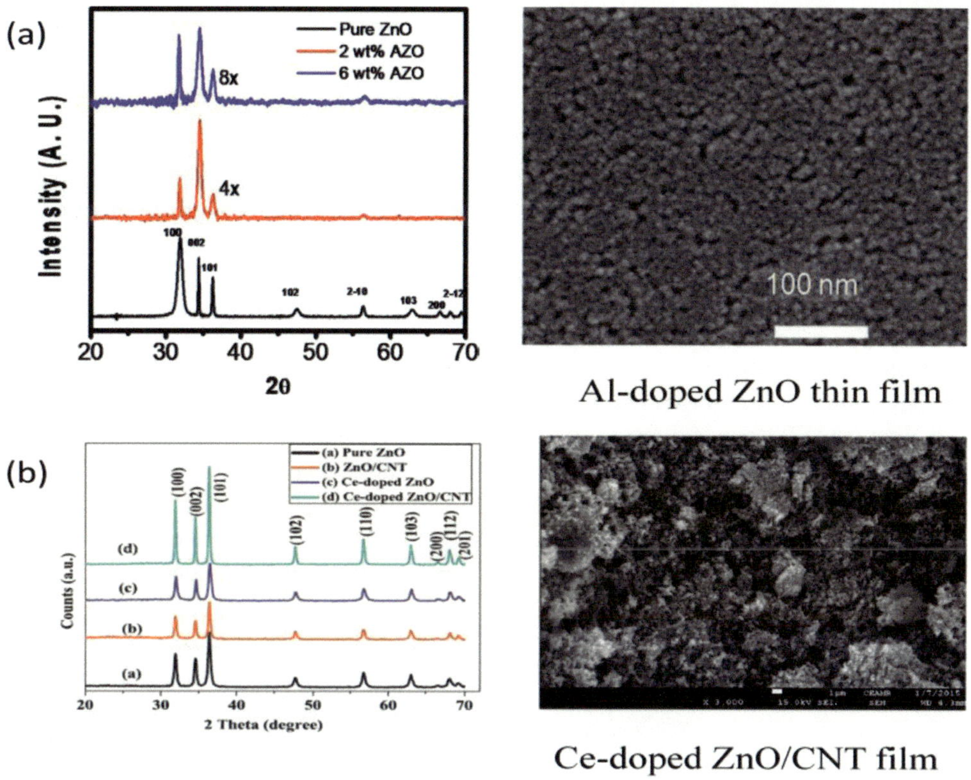

Figure 16.
Characterizations of Al-doped ZnO thin film (a) and Ce-doped ZnO/CNT thin film (b) [49, 50].

Another example is Ce-doped ZnO/carbon nanotube (CNT) thin film, which was prepared using a sol–gel drop coating method [50]. TEA and zinc acetate dehydrate were added into anhydrous ethanol under vigorous stirring. CNT and cerium sulfate were added into the above solution, and the solution was sonicated to get a uniform dispersion. The solution was heated at (55 ± 5)°C with constant stirring at 200 rpm to get the ZnO/CNT sol, which was transparent and quite stable. The ZnO/CNT sol aged for 24 h before the preparation of the Ce-doped ZnO/CNT film. To synthesize Ce-doped ZnO/CNT thin film, the precursor solution was given dropwise on each of that frosted glass substrate using a pipette and then heated for 20 min at 80°C. The coating process was repeated twice for a thick film ($\sim$4 mm) preparation, and finally, the film was annealed at 500°C for 2 h using a muffle furnace.

To evaluate the crystallinity and crystal phases of the prepared thin film, XRD and FESEM spectra were acquired and are displayed in **Figure 16(b)** [50]. Ce and CeO_2, Ce_2O_3 or other crystalline forms were not identified in the patterns, so the authors believed that Ce ions were uniformly substituted by Zn sites or interstitial sites in ZnO crystals. The addition of dopant and functionalized CNTs on the ZnO matrix presented the sharp and high-intense peak indicating the well-crystalline phase of ZnO. From the FESEM spectra of Ce-doped ZnO/CNT thin film, it is observed that that the sample ZnO/CNT is ZnO decorated on the surface of CNT, and the Ce-doped ZnO sample presents 3D flower-like microstructures consisting of many nanoflakes. It is worth noting that many different-size pores were generated in the 3D microstructures, which might act as transport paths for molecules and active sites for catalytic reaction and improve the catalytic properties of the thin-film material [50].

4. Heterogeneous catalysts for wastewater treatment: the process and mechanism

4.1 Heterogeneous Fenton catalysis

Heterogeneous Fenton catalysis, as one of the advanced oxidation technologies, is of great interest for environmental remediation owing to its benign process and general applicability [51]. Heterogeneous Fenton-like reactions on solid catalysts can effectively catalyze the oxidation of organic pollutants at wide pH conditions, which is beneficial for in situ remediations of polluted groundwater and soil and can be reused for further runs. Heterogeneous Fenton-like reaction is a surface-controlled reaction that depends on the catalyst surface area, on H_2O_2 concentration, on the reaction temperature, and on solution pH and ionic strength [52]. Kwan and Voelker [53] have suggested the chain of reactions in the catalyzed oxidation system as follows:

$$\equiv Fe(III) + H_2O_2 \rightarrow \equiv Fe(HO_2)^{2+} + H^+ \quad (1)$$

$$\equiv Fe(HO_2)^{2+} \rightarrow \equiv Fe(II) + \cdot HO_2 \quad (2)$$

$$\equiv Fe(II) + H_2O_2 \rightarrow \equiv Fe(III) + OH - + \cdot OH \quad (3)$$

$$\cdot OH + H_2O_2 \rightarrow H_2O + \cdot HO_2 \quad (4)$$

$$\equiv Fe(II) + \cdot {O_2}^- \rightarrow \equiv Fe(III) + O_2 \quad (5)$$

$$\equiv Fe(III) + \cdot HO_2 \rightarrow \equiv Fe(II) + {HO_2}^- \quad (6)$$

$$\equiv Fe(II) + {HO_2}^- \rightarrow \equiv Fe(III) + \cdot HO_2 \quad (7)$$

If only Fe(III) is originally present, Fe(II) is slowly generated by reactions (1) and (2) initiating oxidation reactions. A simple schematic of heterogeneous Fenton catalysis processes can be seen in **Figure 17**.

The magnetic Fe_3O_4@β-CD/MWCNT nanocomposites were prepared by a hydrothermal method and used as a heterogeneous Fenton-like catalyst to degrade Tetrabromobisphenol A (TBBPA) [54]. Fe_3O_4@β-CD/MWCNT presented good catalytic performance when removing TBBPA from water and got about 97% removal rate of under the optimum condition. Comodification of MWCNT and

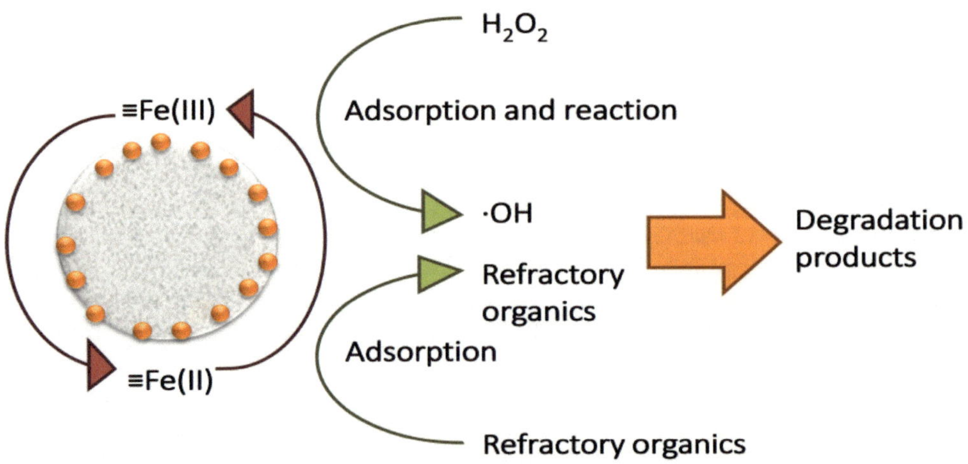

Figure 17.
A simple schematic of heterogeneous Fenton catalysis processes.

β-CD could distinctly promote the removal efficiency of TBBPA, which certified the existence of the synergistic effect between Fe_3O_4, β-CD, and MWCNT. The catalyst has good regeneration ability; it still possessed excellent catalytic activity of TBBPA removal rate at about 86.2% after three cycles. Debromination and bridge fracture between two benzene rings were the primary process of TBBPA mineralization. This catalyst can provide a potential application in the degradation of refractory pollutants [54].

Except for Fe, Cu also can act as an active component of heterogeneous Fenton catalysts. Cu-V bimetallic catalyst (CuVOx) with abundant surface defects was successfully synthesized by the hydrothermal method [55]. Degradation experiment of emerging pollutant fluconazole confirmed that CuVOx exhibited higher catalytic activity, broader pH applicability, and more satisfactory reusability than monometallic copper compounds. By the introduction of vanadium into copper-based materials, several surface properties that are closely related to the reactivity of the copper catalysts, such as adsorption capacity, surface defects, and active site concentration, were improved. The electron-rich center around Cu/V active sites and surface oxygen vacancies are responsible for the rapid dissociation of H_2O_2 and the efficient catalytic oxidation of fluconazole in CuVOx Fenton system. The introduction of vanadium increased active sites and electron density and improved the surface chemical properties of the bimetallic catalyst. FLC and H_2O_2 were firstly adsorbed on the surface of CuVOx, which was beneficial to shorten the mass transfer process of pollutant oxidation. Around oxygen vacancies, there are abundant electrons with high transfer capability, which weaken the O-O bond of adsorbed H_2O_2 and transfer from the material surface to H_2O_2. Under the reaction of catalyst, the adsorbed H_2O_2 was rapidly dissociated to produce a considerable amount of ·OH. Then, ·OH attacked FLC adjacent to the active site, thereby achieving oxidative degradation of the pollutant [55].

4.2 Heterogeneous photocatalysis

Photocatalysis is the process of light-induced redox reactions upon the surrounding molecules to produce radical species for the subsequent utilization in

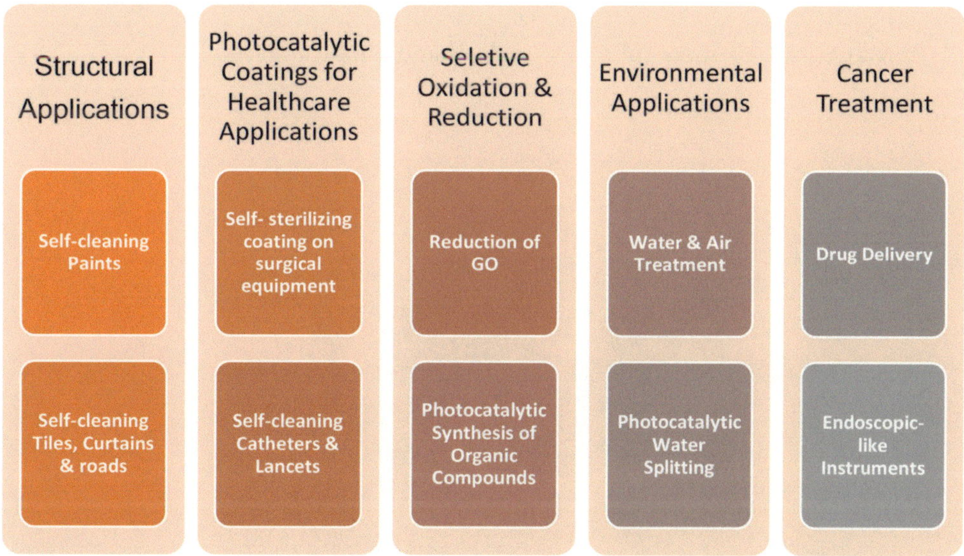

Figure 18.
Applications of heterogeneous photocatalysis [57].

various pollutant degradations, CO_2 conversion into hydrocarbon fuels, and disinfection of microorganisms, and so on, [56] as shown in **Figure 18**. Photocatalysis has many advantages: low reaction temperature, high oxidizability, complete purification, and using solar as an energy source. Furthermore, the careful selection of materials and structures of photocatalysts for photocatalysis is essential to perform the photocatalytic reactions efficiently. An ideal photocatalyst is bound to possess good performances such as narrow band gap energy, suitable band edge potential, reduced recombination, enhanced charge separation, and improved charge transportations [58]. For this purpose, photocatalytic materials are modified for their chemical compositions, which can be done through doping [59], composite formation [60], metal sensitization, and molecule functionalizations [61], among others, and physical structures, which include the size, shape, and surface morphology modifications of the materials.

The principle of photocatalysis is based on the theory of a solid energy band. When the energy of light irradiation absorbed by the semiconductor catalyst is larger than the photon band gap width, the electron–hole pairs are generated due to the transition of electrons. The light irradiation on the semiconductor causes the excitation of electrons from the valence band (VB) to the conduction band (CB) and the creation of holes in the valence band (VB) [62]. Electrons and holes stimulated by light radiation will migrate to the surface of semiconductor particles after various interactions and react with water or organics adsorbed on the surface of semiconductor catalyst particles to produce a photocatalytic effect, as shown in **Figure 19**. The typical photocatalytic process toward pollutant degradation can be described as follows.

$$SC + hv \rightarrow e^- + h^+ \tag{8}$$

$$e^- + O_2 \rightarrow \cdot O_2 \tag{9}$$

$$h^+ + OH^- \rightarrow \cdot OH \tag{10}$$

$$h^+ + H_2O \rightarrow \cdot OH + H^+ \tag{11}$$

$$\cdot O_2 + \text{Org.pollutant} \rightarrow \text{Deg.Products} \tag{12}$$

$$\cdot OH + \text{Org.pollutant} \rightarrow \text{Deg.Products} \tag{13}$$

Here are some examples of photocatalysis and photocatalysts.

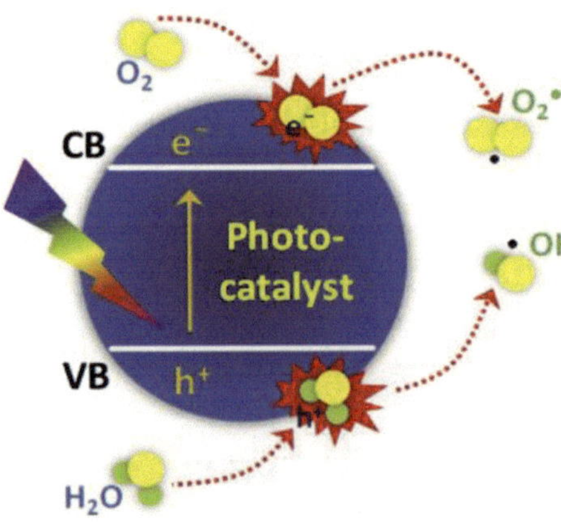

Figure 19.
Photocatalytic process in semiconductor.

A series of TiO_2/biochar composite catalysts were prepared by the hydrolysis method for the degradation of methyl orange, where biochar was obtained from the pyrolysis of waste walnut shells [63]. The photocatalytic activity results showed that TiO_2/biochar composite catalysts exhibited higher catalytic activity than that of pure TiO_2. The decolorization efficiency of 96.88% and the mineralization efficiency of 83.23% were obtained, attributed to the synergistic effect of biochar and TiO_2. After repeated uses, the catalyst still exhibited rather high activity toward the degradation of MO, where the decolorization efficiency and mineralization efficiency of MO achieved 92.45 and 76.56%, and the loss of activity was negligible. The possible degradation mechanism of MO with TiO_2/biochar composites can be explained as follows. MO molecules firstly adsorbed by biochar due to the well-developed pore structure, which provides more opportunities for the contact of catalysts and pollutants. With irradiation of ultraviolet light, TiO_2 was excited to generate electrons from the valence band, transferring to the conduction band, and holes were left in the valence band. Biochar transferred electrons and acted as acceptors, inhibiting the recombination of electron–hole pairs. Then, the electrons would react with oxygen molecules and water molecules to generate superoxide anion radicals and hydroxyl radicals. Holes would also be trapped by water molecules to engender hydroxyl radicals. Thus, holes, superoxide anion radicals, and hydroxyl radicals cofunctioned as active species would interact with the MO molecules, and finally, carbon dioxide was formed [63].

Sun et al. evaluated the photocatalytic efficiency of activated carbon-supported TiO_2 catalyst (AC/TiO_2) for degradation of aflatoxin B1 (AFB1) under UV–Vis light. AC/TiO_2 was prepared by simple hydrothermal synthesis [64]. According to the experiments, the higher degradation efficiency of AFB1 by AC/TiO_2 composite (98%) than that by bare TiO_2 (76%) was attributed to a higher surface area of AC/TiO_2 and enhanced visible-light intensity by the synergistic effect of TiO_2 and AC. Moreover, the catalyst can be easily separated from the solution and can still keep high activity. The authors found that the hole (h+) and the hydroxyl radicals (·OH) played an essential role in the degradation of AFB1 and proposed the possible photocatalytic oxidation mechanism of AC/TiO_2 for AFB1 degradation. When irradiated by UV light, the excited AC/TiO_2 composite produces electrons (e−) to the conduction band and holes (h+) on the valence band. The photo-generated electrons can react with O_2 to produce superoxide radicals ($O_2^{-}\cdot$) and further produce hydroxyl radicals (·OH). The photo-generated holes can react with hydroxide ions (OH^-) in the solution to directly produce ·OH. Some of these holes (h+) and ·OH would oxidize AFB1, which is absorbed on the surface of the catalyst [64]. Most of the photocatalysis processes have the same mechanism and they can be illustrated by (**Figure 20**).

4.3 Heterogeneous electrocatalysis

Electrocatalysis is a kind of catalysis that accelerates the charge transfer reaction at the interface of electrodes and electrolytes. The range of electrode catalysts is limited to electrical materials such as metals and semiconductors. There are many kinds of semiconductor oxides, such as skeleton nickel, nickel boride, tungsten carbide, sodium tungsten bronze, spinel and tungsten minerals, as well as various metals and phthalocyanines. Electrocatalysis is mainly used in the treatment of organic wastewater, degradation of chromium-containing wastewater, desulfurization of flue gas and raw coal, the simultaneous removal of NOx and SO_2, and the reduction of carbon dioxide and nitrogen. The selection of appropriate electrode materials plays a critical role in accelerating the electrode reaction. The selected electrode materials play a catalytic role in the process of electrification, thus

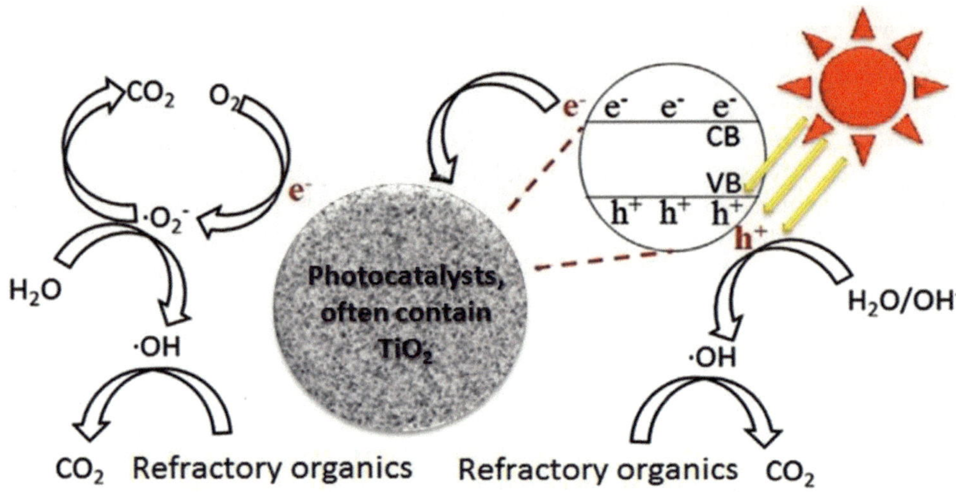

Figure 20.
The schematic of photocatalysis processes.

changing the reaction rate or direction of the electrode without any qualitative change in itself. Electrocatalysis covers two aspects of electrode reaction and catalysis, so electrocatalysts must have two functions at the same time: conductive and relatively free transfer of electrons; and effective catalytic activation. Therefore, the feasible way to design electrocatalysts is to modify the electrodes. Combining active components with conductive base electrodes in the form of covalent bonds or chemical adsorption can achieve the dual purposes of both transferring electrons and activating substrates.

Chao Zhang et al. investigated modified iron-carbon with polytetrafluoroethylene (PTFE) as a heterogeneous electro-Fenton (EF) catalyst for 2,4-dichlorophenol (2,4-DCP) degradation in near-neutral pH condition [65]. The degradation performance of the catalyst modification with 20% PTFE maintained well with low iron leaching. Moreover, the degradation efficiency of 2,4-DCP could exceed 95% within 120 min treatment at catalyst dosage of 6 g/L, initial pH of 6.7, and the current intensity of 100 mA. Two-stages of 2, 4-DCP degradation were observed: the first stage is a slow anodic oxidation stage, and the second one is a much faster heterogeneous EF oxidation stage. The automatic decrease of pH in the first stage initiated the Fe^{2+} release from microelectrolysis and benefited to the subsequent EF reaction. Aromatic intermediates including 2-chlorohydroquinone, 4,6-dichlororesorcinol, and 3,5-dichlorocatechol were detected by GC–MS. A possible mechanism and degradation pathway for 2,4-DCP were proposed. Air permeated from the prepared ADE cathode and was utilized to produce H_2O_2. Simultaneously, Fe-C microelectrolysis would produce Fe^{2+}, which reacted with H_2O_2 as Fenton reagent. Considering the results of GC–MS and IC analyses, 2,4-DCP can be hydroxylated by ·OH addition's reaction onto the aromatic ring, resulting in the formation of 3,5-dichlorocatechol and 4,6-dichlororesorcinol. Meanwhile, the chlorine atom located in the paraposition on the aromatic ring was substituted by ·OH to yield 2-chlorohydroquinone. The aromatic organics mentioned above would be further oxidized to organic acids by ·OH, such as oxalic, acetic, and formic acids until mineralized to CO_2 completely. The possible degradation paths of 2,4-DCP can be seen in **Figure 21**.

Nanolayered double hydroxide (NLDH) decorated with Fe and Cu was applied as a novel heterogeneous catalyst for catalytic degradation of gentamicin by the electro-Fenton (EF) process [66]. The EF process was equipped with a graphite plate under aeration. The highest removal efficiency was 91.3% when the Cu-Fe--NLDH-equipped EF process applied in comparison with the Fenton (50%) and the electro-oxidation alone (25.6%). Increasing the current resulted in the enhanced degradation of gentamicin, while the excessive electrolyte concentration and catalyst dosage led to the tangible drop in the reactor performance. At a specified reaction time, the injection of O_3 gas enhanced the efficiency of the Cu-Fe-NLDH-equipped EF process. The presence of ethanol led to a more suppressing effect than benzoquinone, indicating the dominant role of ·OH radical in the degradation of gentamicin compared with other free radical species such as O_2^{-}· radical. Most of

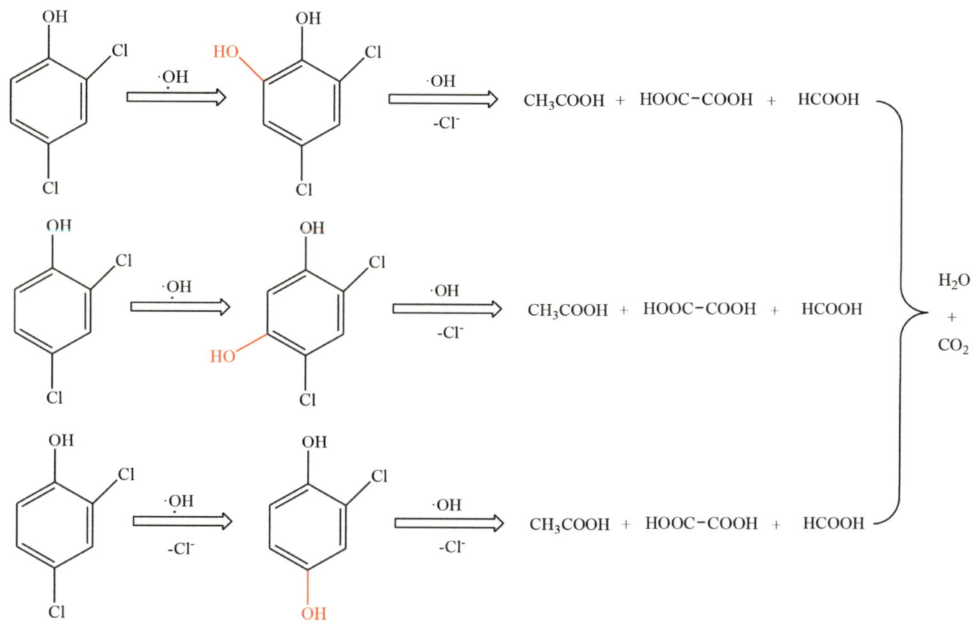

Figure 21.
Proposed degradation pathway of 2,4-DCP [65].

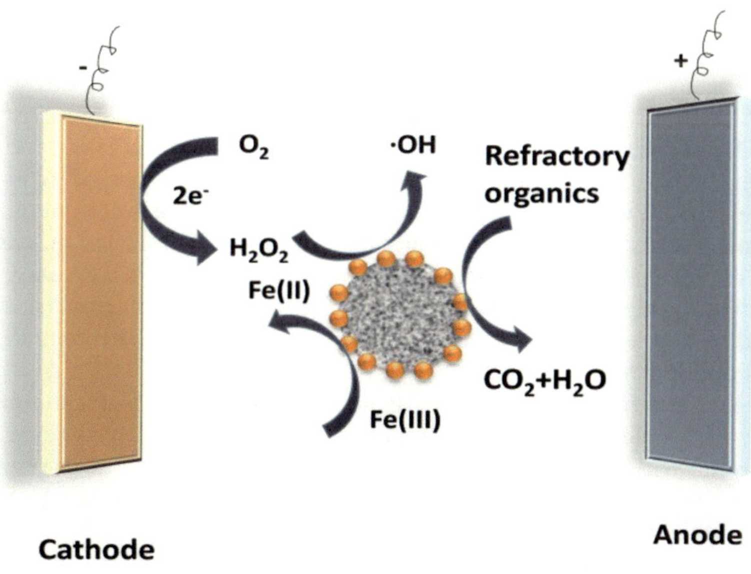

Figure 22.
The schematic of electrocatalysis processes [66].

Catalysts	Target pollutants	Assistant	Degradation efficiency	Reference
Fe/OATP	HA-Na	—	97%	[24]
Fe_3O_4@β-CD/ MWCNT	Tetrabromobisphenol A	—	97%	[54]
CuVOx	Fluconazole	—	99%	[55]
Fe_2O_3/ATP	Methylene blue	—	99%	[67]
CuO/SiO_2	RhB	—	95%	[68]
TiO_2/biochar	Methyl orange	UV–Vis light	98%	[63]
AC/TiO_2	Aflatoxin B1	UV–Vis light	96.88%	[64]
Mg-ZnO-Al_2O_3	Caffeine	UV–Vis light	98.9%	[69]
Fe-C/PTFE	2,4-DCP	Electric	95%	[65]
Cu-Fe-NLDH	Gentamicin	Electric	91.3%	[66]
DG@Fe_3O_4	Methylene blue	Electric	99%	[70]

Table 4.
Comparison of some catalysts for different pollutant degradation.

the photocatalysis processes have the same mechanism, they just used different electrodes and catalysts, and they can be illustrated by (**Figure 22**).

The comparison of some catalysts for different pollutant degradation was shown in **Table 4**.

5. Conclusions

Heterogeneous catalytic processes, as an efficient green method coping with organic wastewater, have attracted considerable attention. Heterogeneous catalysts, such as bulk catalysts and supported catalysts (including surface-loaded catalysts, core-shell catalysts, and thin-film catalysts), are often synthesized by impregnation, sol–gel, and coprecipitation methods. To characterize heterogeneous catalysts, BET surface area, scanning electron microscope (SEM), transmission electron microscope (TEM), Fourier transform infrared spectroscopy (FT-IR), and X-ray diffraction (XRD) are common characterizing methods. Heterogeneous catalytic processes are efficient in treating organic wastewater due to the generation of highly reactive and nonselective hydroxyl radicals (·OH), which can oxidize and mineralize most organic compounds, with the advantages of no secondary pollution, easy separation, and recycling. The researchers often employed heterogeneous catalysts to degrade dyes, pesticides, and antibiotics, among others. The reactions and mechanisms of different heterogeneous catalytic processes might be different according to the target pollutants.

At present, the researches on heterogeneous-supported catalysts should focus on the following problems: (1) how to prepare heterogeneous catalysts (especially nanocatalysts) with high efficiency, low cost, and low energy consumption, which can be widely used in the treatment of industrial wastewater; (2) how to prepare the catalysts with larger specific surface area and higher activity, and how to make the loaded active components distribute more evenly on the surface of the carrier; and (3) how to expand the application fields of heterogeneous catalysts, such as the application in degradation of microplastics in wastewater.

Author details

Ting Zhang
School of Petrochemical Engineering, Lanzhou University of Technology, Lanzhou, China

*Address all correspondence to: zhangting@lut.cn

IntechOpen

© 2020 The Author(s). Licensee IntechOpen. Distributed under the terms of the Creative Commons Attribution - NonCommercial 4.0 License (https://creativecommons.org/licenses/by-nc/4.0/), which permits use, distribution and reproduction for non-commercial purposes, provided the original is properly cited.

References

[1] Taylor HS. Catalysis—Chemical Process. 1957. Available from: www.britannica.com

[2] Wang Q, Tian S, Long J, Ning P. Use of Fe(II)Fe(III)-LDHs prepared by co-precipitationmethod in a heterogeneous-Fenton process for degradation of methylene blue. Catalysis Today. 2014:41-48

[3] Cheng M, Ma W, Li J, Huang Y, Zhao J, Wen YX, et al. Visible-light-assisted degradation of dye pollutants over Fe(III)-loaded resin in the presence of H_2O_2 at neutral pH values. Environmental Science & Technology. 2004;**38**:1569-1575

[4] Socci J, Osatiashtiani A, Kyriakou G, Bridgwater T. The catalytic cracking of sterically challenging plastic feedstocks over high acid density Al-SBA-15 catalysts. Applied Catalysis A: General. 2019;**570**:218-227

[5] Lu P, Cai F, Zhang J, Liu Y, Sun Y. Hydrogen production from methanol steam reforming over B-modified $CuZnAlO_x$ catalysts. Journal of Fuel Chemistry and Technology. 2019;**47**: 791-798

[6] Cai F, Lu P, Ibrahim JJ, Fu Y, Zhang J, Sun Y. Investigation of the role of Nb on Pd−Zr−Zn catalyst in methanol steam reforming for hydrogen production. International Journal of Hydrogen Energy. 2019;**44**:11717-11733

[7] Islam M, Basnayake R, Korzeniewski C. A study of formaldehyde formation during methanol oxidation over PtRu bulk alloys and nanometer scale catalyst. Journal of Electroanalytical Chemistry. 2007;**599**:31-40

[8] Ji F, Li C, Zhang J, Deng L. Heterogeneous photo-Fenton decolorization of methylene blue over $LiFe(WO_4)_2$ catalyst. Journal of Hazardous Materials. 2011;**186**: 1979-1984

[9] Duan F, Yang Y, Li Y, Cao H, Wang Y, Zhang Y. Heterogeneous Fenton-like degradation of 4-chlorophenol using iron/ordered mesoporous carbon catalyst. Journal of Environmental Sciences. 2014; **26**:1171-1179

[10] Hu X, Liu B, Deng Y, Chen H, Luo S, Sun C, et al. Adsorption and heterogeneous Fenton degradation of 17α-methyltestosterone on nano Fe_3O_4/MWCNTs in aqueous solution. Applied Catalysis B: Environmental. 2011;**107**: 274-283

[11] Molina R, Martínez F, Melero JA, Bremner DH, Chakinala AG. Mineralization of phenol by a heterogeneous ultrasound/Fe-SBA-15/H_2O_2 process: Multivariate study by factorial design of experiments. Applied Catalysis B: Environmental. 2006;**66**: 198-207

[12] Muthukumari B, Selvam K, Muthuvel I, Swaminathan M. Photoassisted hetero-Fento mineralization of azo dyes by Fe(II)-Al_2O_3 catalyst. Chemical Engineering Journal. 2009;**153**:9-15

[13] Vignesh K, Rajarajan M, Suganthi A. Visible light assisted photocatalystic performance of Ni and Th co-doped ZnO nanoparticles for the degradation of methylene blue dye. Journal of Industrial and Engineering Chemistry. 2014;**20**:3826-3833

[14] Panda N, Sahoo H, Mohapatra S. Decolourization of methyl orange using Fentonlike mesoporous Fe_2O_3–SiO_2 composite. Journal of Hazardous Materials. 2011;**185**:359-365

[15] Zhong X, Royer S, Zhang H, Huang Q, Xiang L, Valange S, et al.

Mesoporous silica iron-doped as stable and efficient heterogeneous catalyst for the degradation of C.I. acid orange 7 using sono–photo-Fenton process. Separation and Purification Technology. 2011;**80**:163-171

[16] Liu T, You H, Chen Q. Heterogeneous photo-Fenton degradation of polyacrylamide in aqueous solution over Fe(III)–SiO_2 catalyst. Journal of Hazardous Materials. 2009;**162**:860-865

[17] Vu AT, Xuan TN, Lee CH. Preparation of mesoporous $Fe_2O_3{\cdot}SiO_2$ composite from rice husk as an efficient heterogeneous Fenton-like catalyst for degradation of organic dyes. Journal of Water Process Engineering. 2019;**28**: 169-180

[18] Chen JX, Zhu LZ. Catalytic degradation of Orange II by UV-Fenton with hydroxyl-Fe-pillared bentonite in water. Chemosphere. 2006;**65**: 1249-1225

[19] Chen JX, Zhu LZ. Oxalate enhanced mechanism of hydroxyl-Fe-pillared bentonite during the degradation of Orange II by UV-Fenton process. Journal of Hazardous Materials. 2011; **185**:1477-1481

[20] Feng JY, Hu XJ, Yue PL. Discoloration and mineralization of Orange II by using a bentonite clay-based Fe nanocomposite film as a heterogeneous photo-Fenton catalyst. Water Research. 2005;**39**:89-96

[21] Perderiset M, Baillif P, Jaurand MC. Chemical analysis and photoelectron spectroscopy of the adsorption of macromolecules on the surface of attapulgite. Journal of Colloid and Interface Science. 1988;**121**:381-391

[22] Zhang J, Chen H, Wang A. Study on superabsorbent composite. III. Swelling behaviors of polyacrylamide/attapulgite composite based on acidified attapulgite and organo-attapulgite. European Polymer Journal. 2005;**41**:2434-2442

[23] Li A, Zhang JP, Wang AQ. Utilization of starch and clay for the preparation of superabsorbent composite. Bioresource Technology. 2007;**98**:327-332

[24] Zhang T, Qian C, Dong L. Preparation of particle reusable heterogeneous catalyst Fe_3O_4/ATP for methylene blue decolorization. International Journal of Environmental Science and Technology. 2019;**10**: 343-349

[25] Zhang T, Liu J, Chen M, Wang Y. Preparation of heterogeneous Fenton catalyst Fe/organo-attapulgite and its performance in sodium humate degradation. Desalination and Water Treatment. 2018;**107**:91-99

[26] Solar P, Polonskyi O, Olbricht A, Hinz A, Shelemin A, Kylián O, et al. Single step generation of metal-plasma polymer multicore@shell nanoparticles from the gas phase. Scientific Reports. 2017;7: 8514-8519

[27] Martinez L, Mayoral A, Espineira M, Roman E, Palomares FJ, Huttel Y. Core@ shell, Au@ TiO_x nanoparticles by gas phase synthesis. Nanoscale. 2017; **9**:6463-6470

[28] Solaiyammal T, Murugakoothan P. Green synthesis of Au and Au@ TiO_2 core-shell structure formation by hydrothermal method for dye sensitized solar cell applications. Journal of Materials Science: Materials in Electronics. 2018;**29**:1-9

[29] Guo XH, Ma JQ, Ge HG. $CoFe_2O_4$@ TiO_2@ Au core-shell structured microspheres: Synthesis and photocatalyltic properties. Russian Journal of Physical Chemistry A. 2017; **91**:2643-2650

[30] Shi BN, Wan JF, Liu CT, Yu XJ, Ma FW. Synthesis of $CoFe_2O_4$/MCM-41/TiO_2 composite microspheres and its performance in degradation of phenol. Materials Science in Semiconductor Processing. 2015;**37**:241-249

[31] Sohail M, Xue H, Jiao Q, Li H, Khan K, Wang S, et al. Synthesis of well-dispersed TiO_2/CNTs@ $CoFe_2O_4$nanocomposites and their photocatalytic properties. Materials Research Bulletin. 2018;**101**:83-89

[32] Hang S, He Q, Ping S, Zeng S, Xu K, Li J, et al. One-pot synthesis of Au@TiO_2 yolk-shell nanoparticles with enhanced photocatalytic activity under visible light. Journal of Colloid and Interface Science. 2017;**505**:884-891

[33] Ma JQ, Guo SB, Guo XH, Ge HG. Liquid-phase deposition of TiO_2nanoparticles on core-shell Fe_3O_4 @SiO_2spheres: Preparation, characterization, and photocatalytic activity. Journal of Nanoparticle Research. 2015;**17**:1-11

[34] Wu J, Yang Y, Ding J. Green production of cyclohexanone oxime with a fixed-bed reactor using shaped titanosilicate catalysts. Chinese Journal. 2015;**60**:1538-1545

[35] Caruso F. Nanoengineering of particle surfaces. Advanced Materials. 2010;**13**:11-19

[36] Reiss P, Protiere M, Li L. Core/shell semiconductor nanocrystals. Small. 2009;**5**:154-168

[37] Shi ZL, Du C, Yao SH. Preparation and photocatalytic activity of cerium dopedanatase titanium dioxide coated magnetite composite. Journal of the Taiwan Institute of Chemical Engineers. 2011;**42**:652-657

[38] Chou KS, Lin MY, Wu HH. Studies on the removal of 2-propanol by Ag@Fe_2O_3 core–shell structured catalyst. Journal of the Taiwan Institute of Chemical Engineers. 2013; **44**:228-232

[39] Jilani A, Abdel-Wahab MS, Hammad AH. Advance Deposition Techniques for Thin Film and Coating. London: IntechOpen; 2017

[40] Shwetharani R, Chandan HR, Sakar M, Balakrishna GR, Reddy KR, Raghu AV. Photocatalytic semiconductor thin films for hydrogen production and environmental applications. International Journal of Hydrogen Energy. 2019. DOI: 10.1016/j.ijhydene.2019.03.149

[41] Arconada N, Suarezb DAS, Portelab R, Coronadob JM, Sanchezb B, Castro Y. Synthesis and photocatalytic properties of dense and porous TiO_2-anatase thin films prepared by solegel. Applied Catalysis B: Environmental. 2009;**86**:1-7

[42] Sakthivel S, Shankar MV, Palanichamy M, Arabindoo B, Bahnemann DW, Murugesan V. Enhancement of photocatalytic activity by metal deposition: Characterization and photonic efficiency of Pt, Au, and Pd deposited on TiO_2 catalyst. Water Research. 2004;**38**:3001-3008

[43] Jung JM, Wang M, Kim EJ, Park C, Hahn SH. Enhanced photocatalytic activity of Aubuffered TiO_2 thin films prepared by radio frequency magnetron sputtering. Applied Catalysis B: Environmental. 2008;**84**:389-392

[44] Seong SG, Kim EJ, Kim YS, Lee KE, Hahn SH. Influence of deposition atmosphere on photocatalytic activity of TiO_2/SiO_x double-layers prepared by RF magnetron sputtering. Applied Surface Science. 2009;**256**:1-5

[45] Di Mauro A, Fragala ME, Privitera V, Impellizzeri G. ZnO for application in photocatalysis: From thin films to nanostructures. Materials

Science in Semiconductor Processing. 2017;**69**:44-51

[46] Chen YH, Tu KJ. Thickness dependent o photocatalytic activity of hematite thin films. International Journal of Photoenergy. 2012; **1110-662X**:1-6

[47] Schultz AM, Salvador PA, Rohrer GS. Enhanced photochemical activity of α-Fe_2O_3 films supported on $SrTiO_3$ substrates under visible light illumination. Chemical Communications. 2012;**48**:2012-2014

[48] Bhachu DS, Sathasivam S, Carmalt CJ, Parkin IP. PbO modified TiO_2 thin films: A route to visible light photocatalysts. Langmuir. 2014;**30**: 624-630

[49] Islam MR, Rahman M, Farhad SFU, Podder J. Structural, optical and photocatalysis properties of sol–gel deposited Aldoped ZnO thin films. Surface and Interface Analysis. 2019;**16**: 120-126

[50] Elias M, Uddin MN, Hossain MA, Saha JK, Siddiquey IA, Sarker DR, et al. An experimental and theoretical study of the effect of Ce doping in ZnO/CNT composite thin film with enhanced visible light photo-catalysis. International Journal of Hydrogen Energy. 2019;**44**:20068-20078

[51] Xia M, Chen C, Long M, Chen C, Cai W, Zhou B. Magnetically separable mesoporous silica nanocomposite and its application in Fenton catalysis. Microporous and Mesoporous Materials. 2011;**145**:217-223

[52] Matta R, Hanna K, Chiron S. Fenton-like oxidation of 2,4,6-trinitrotoluene using different iron minerals. Science of the Total Environment. 2007;**385**:242-251

[53] Kwan WP, Voelker BM. Rates of hydroxyl radical generation and organic compound oxidation in mineral-catalyzed Fenton-like systems. Environmental Science & Technology. 2005;**39**:9303-9308

[54] Zhang Y, Wu P, Chen Z, Zhou L, Zhao Y, Lai Y, et al. Synergistic effect in heterogeneous Fenton degradation of tetrabromobisphenol a by MWCNT and β-CD co-modified Fe_3O_4. Materials Research Bulletin. 2019;**113**:14-24

[55] Zhang N, Xue C, Wang K, Fang Z. Efficient oxidative degradation of fluconazole by a heterogeneous Fenton process with Cu-V bimetallic catalysts. Chemical Engineering Journal. 2020; **380**:122516

[56] Francis O, Kuben GK, Sittert CGCE, Poomani GP. Recent progress in the development of semiconductor-based photocatalyst materials for applications in photocatalytic water splitting and degradation of pollutants. Advanced Sustainable Systems. 2017: 1700006

[57] Ahmed SN, Haider W. Heterogeneous photocatalysis and its potential applications in water and wastewater treatment: A review. Nanotechnology. 2018;**29**: 342001

[58] Reddy KR, Reddy CV, Nadagouda MN, Shetti NP, Jaesool S, Aminabhavi TM. Polymeric graphitic carbon nitride (g-C_3N_4)-based semiconducting nanostructured materials: Synthesis methods, properties and photocatalytic applications. Journal of Environmental Management. 2019; **238**:25-40

[59] Ravikumar CH, Mahto MSA, Nanjundaiah RT, Thippeswamy R, Teixeira SR, Balakrishna RG. Observation of oxo-bridged yttrium in TiO_2 nanostructures and their enhanced photocatalytic hydrogen generation under UV/visible light irradiations.

Materials Research Bulletin. 2018;**104**: 212-219

[60] D'Souza LP, Muralikrishna S, Chandan HR, Ramakrishnappa T, Balakrishna RG. Neodymium doped titania as photoanode and graphene oxide-CuS composite as a counter electrode material in quantum dot solar cell. Journal of Materials Research. 2015; **30**:3241-3251

[61] Schiffman HRCJD, Balakrishna RG. Quantum dots as fluorescent probes: Synthesis, surface chemistry, energy transfer mechanisms, and applications. Sensors and Actuators B: Chemical. 2018;**258**:1191-1214

[62] Ibhadon A, Fitzpatrick P. Heterogeneous photocatalysis: Recent advances and applications. Catalysts. 2013;**3**:189

[63] Lu L, Shan R, Shi Y, Wang S, Yuan H. A novel TiO_2/biochar composite catalysts for photocatalytic degradation of methyl orange. Chemosphere. 2019;**222**:391-398

[64] Sun S, Zhao R, Xie Y, Liu Y. Photocatalytic degradation of aflatoxin B1 by activated carbon supported TiO_2 catalyst. Food Control. 2019;**100**: 183-188

[65] Zhang C, Zhou M, Ren G, Yu X, Ma L, Yang J, et al. Heterogeneous electro-Fenton using modified iron-carbon as catalyst for 2,4-dichlorophenol degradation: Influence factors, mechanism and degradation pathway. Water Research. 2015;**70**: 414-424

[66] Ghasemi M, Khataee A, Gholami P, Soltani RDC. Template-free microspheres decorated with Cu-Fe-NLDH for catalytic removal of gentamicin in heterogeneous electro-Fenton process. Journal of Environmental Management. 2019;**248**: 109236

[67] Zhang T, Chen M, Yu S. Decolorization of methylene blue by an attapulgite-based heterogeneous Fenton catalyst: Process optimisation. Desalination and Water Treatment. 2017;**63**:275-282

[68] Sun Y, Tian P, Ding D, Yang Z, Wang W, Xin H, et al. Revealing the active species of Cu-based catalysts for heterogeneous Fentonreaction. Applied Catalysis B: Environmental. 2019;**258**: 117985

[69] Elhalil A, Elmoubarki R, Farnane M, Machrouhi A, Sadiq M, Mahjoubi FZ, et al. Photocatalytic degradation of caffeine as a model pharmaceutical pollutant on Mg doped ZnO-Al_2O_3heterostructure. Environmental Nanotechnology, Monitoring and Management. 2018;**10**:63-72

[70] Wang J, Yang F, Wang S, Zhong H, Wu Z, Cao Z. Reactivation of nano-Fe_3O_4/diethanolamine/rGO catalyst by using electric field in Fenton reaction. Journal of the Taiwan Institute of Chemical Engineers. 2019;**99**:113-122

Chapter 3

Applications of Chemical Kinetics in Heterogeneous Catalysis

Zhenhua Zhang, Li-Ping Fan and Yue-Juan Wang

Abstract

Chemical kinetics is a key subdiscipline of physical chemistry that studies the reaction rate in every elemental step and corresponding catalytic mechanism. It mainly concludes molecular reaction dynamics, catalytic dynamics, elemental reaction dynamics, macrodynamics, and microdynamics. Such a research field has wide applications in heterogeneous catalysis. Based on the Arrhenius plot fitted by the catalytic conversions below 15% without the mass transfer effect and heat transfer effect, the apparent activation energy echoing with the intrinsically catalytic sites and the pre-exponential factor echoing with the relative number of active sites can be, respectively, derived from the slope and intercept of the Arrhenius plots, which can be used to compare the intrinsically catalytic activity of different catalysts and the relative amount of active sites. Reaction orders of both reactants and products are derived from the reaction rate equation and also fitted by the catalytic conversions below 15% without the mass transfer effect and heat transfer effect. According to the acquired reaction orders, the reaction mechanism can be proposed and even defined in some simple reactions. Therefore, investigations of chemical kinetics are of extreme importance and meaning in heterogeneous catalysis.

Keywords: Arrhenius equation, reaction rate equation, apparent activation energy, reaction order, heterogeneous catalysis

1. Introduction

Chemical kinetics, also known as reaction kinetics or chemical reaction kinetics, is a key branch of physical chemistry. Its main task is to investigate the rate during chemical process and to propose the catalytic mechanism of chemical process by a method of research object acting as a nonequilibrium dynamic system whose properties change with time [1, 2]. Through the study of chemical kinetics, it can reasonably guide us to know how to control the reaction conditions and improve the main reaction rate, in order to increase the production of chemical products, and also guide us to learn how to suppress or slow down the reaction rate of side reactions to reduce the consumption of raw materials, reduce the burden of separation operations, and eventually improve the product quality. Chemical kinetics can provide general knowledge on how to avoid explosion of dangerous goods, material corrosion, and aging and deterioration of products. It can also carry out optimal design and control for the industrialization of scientific research results and select the most suitable operating conditions for the existing production. Generally speaking, chemical kinetics is one of the main theoretical bases of chemical reaction engineering, accompanied by the chemical thermodynamics for the improved study of chemical reactions.

Chemical kinetics is of great antiquity with continuous improvements. Date back to the first half of the twentieth century, a great deal of research was devoted to the parameter determination, theoretical analysis, and using these parameters to study the reaction mechanism. However, investigation of the reaction mechanism was extremely difficult at that time, which was dominantly limited by the abilities of probe and analyze reaction intermediates. As time goes on, the study of free radical chain reaction kinetics was generally carried out in the late twentieth century, which brought two development trends for chemical kinetics. One was for the study of elemental reaction dynamics, and the other was to urgently establish a method for detecting active intermediates. Such researches were accompanied by the development of electronics and laser technology. Notably, Herschbach, Lee, and Polanyi, three famous scientists in this field, were awarded the Nobel Prize in chemistry in 1986, which marked the importance of chemical kinetics and indicated the current progress and the achieved level.

Nowadays, chemical kinetics has become an indispensable tool in both chemical discipline development and scientific research. The application in heterogeneous catalysis is a typical example [3]. Heterogeneous catalysis, possessing most of the catalytic reactions in the industry, refers to the catalytic reaction that occurs at the interface of two phases. The heterogeneously catalytic reaction generally occurs on the catalyst surface on which the reactant molecular adsorbs, diffuses, activates, reacts, and desorbs to acquire the final products. Therefore, the surface compositions and structures of catalytic materials are extremely critical in determining the catalytic properties [4]. The adsorption center on catalyst surface is often the center of catalytic activity, which is denoted as active site. The reactant molecule bonds to the active site to form surface adsorption complex, which is denoted as active intermediate species. The existence of active site on catalyst surface can affect the formation and activation of active intermediate species, thus altering the reaction path and consequently the required activation energy [5, 6].

In this chapter, two important kinetic equations, the Arrhenius equation and the reaction rate equation, were comprehensively introduced, accompanied by the applications in heterogeneous catalysis. Through the introduction of the Arrhenius equation, the apparent activation energy (E_a) and pre-exponential factor (A) can be calculated based on the Arrhenius plots fitted by the catalytic conversions below 15% without the mass transfer effect and heat transfer effect, which can reflect the intrinsically catalytic sites and relative number of active sites, respectively. The reaction mechanism involving rate-determining elementary reaction step can be proposed by the reaction order derived from reaction rate equation plotted by the catalytic conversions below 15% without the mass transfer effect and heat transfer effect. Understanding the structure and number of active sites and catalytic mechanism, it can guide us insights into the understanding for a catalytic reaction and thereby provide a research fundamental for the design and synthesis of high-performance catalysts.

2. Applications of kinetic equations in heterogeneous catalysis

2.1 The Arrhenius equation and its applications in heterogeneous catalysis

The Arrhenius equation is an empirical formula with chemical reaction rate changing with temperature, which was established by Svante August Arrhenius, a famous scientist in Sweden. This equation can be expressed in different forms containing the differential expression (Eq. (1)), antiderivative form expression (Eq. (2)), derivative form expression (Eq. (3)), and exponential form expression (Eq. (4)), as listed below. Among them, k represents the rate constant, R represents the gas molar constant, T represents the thermodynamic temperature, E_a represents the apparent activation energy, and A represents the pre-exponential factor. The Arrhenius equation can be used to calculate the apparent activation energy, the optimal reaction temperature and time, and the reaction rate constant in a chemical reaction.

$$\frac{dlnk}{dT} = \frac{E_a}{RT^2} \tag{1}$$

$$lnk = \frac{-E_a}{RT} + lnA \tag{2}$$

$$ln\frac{k_2}{k_2} = \frac{-E_a}{R}\left(\frac{1}{T_2} - \frac{1}{T_1}\right) \tag{3}$$

$$k = e^{\frac{-E_a}{RT}} \tag{4}$$

As shown in the above listed equations, it shows a wide applicability in catalytic reactions, including not only gas-phase reactions but also liquid-phase reactions, as well as most of complex catalytic reactions. However, the precondition of this equation in use is to assume that E_a is an independent constant without relating to temperature. Therefore, the results derived from this equation is well agreeing with the real experiment results within a certain temperature range.

This equation is of wide applications in heterogeneous catalysis. Typical in the catalytic oxidation of CO, a representative probe reaction in heterogeneous catalysis [7–12], the Arrhenius equation is often used as an evaluation standard to compare the intrinsic activity of different catalysts and relative amount of active sites. Bao et al. [7] reported a crystal-plane-controlled surface restructuring and catalytic performance of Cu_2O nanocrystals in CO oxidation with excess O_2. As shown in **Figure 1**, the relationships of catalyst structure and catalytic property were deeply explored. Based on the microscopic and spectroscopic characterization results (detailed descriptions shown in Ref. [7]), the surface compositions and

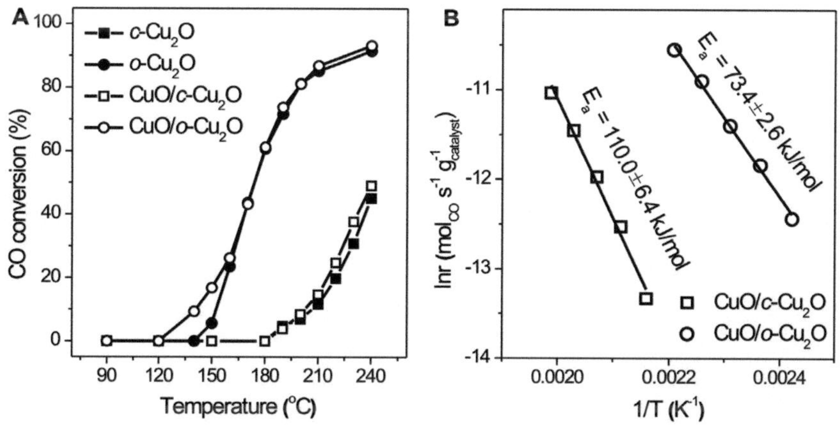

Figure 1.
(A) Catalytic performance of c-Cu_2O, o-Cu_2O, CuO/c-Cu_2O, and CuO/o-Cu_2O in CO oxidation reaction and (B) the Arrhenius plot of CO oxidation catalyzed by CuO/o-Cu_2O and CuO/c-Cu_2O [7]. Copyright © 2011, Wiley-VCH Verlag GmbH & Co.

structures of the restructured CuO/Cu_2O catalysts were distinctly identified. Their corresponding catalytic performance in CO oxidation with the plotted Arrhenius equation was also conducted. The derived E_a value plays a key role in comparing the intrinsic activity of surface sites on different catalysts based on the slope of the Arrhenius plots. CuO/Cu_2O octahedra (denoted as $CuO/o\text{-}Cu_2O$) show a lower E_a value of 73.4 ± 2.6 kJ mol^{-1} than CuO/Cu_2O cubes (denoted as $CuO/c\text{-}Cu_2O$) with a E_a value of 110.0 ± 6.4 kJ mol^{-1}, indicating that CO oxidation catalyzed by the CuO of $CuO/o\text{-}Cu_2O$ surface is more intrinsically active than the CO oxidation catalyzed by the CuO of $CuO/c\text{-}Cu_2O$ surface. Density functional theory (DFT) calculations results demonstrate that they exist in different catalytic reaction mechanisms involved in different CuO/Cu_2O surface structures. The $CuO/Cu_2O(100)$ surface is terminated with two-coordinated oxygen (O_{2c}) atoms, which result in a typical Mars-Van Krevelen (MvK) mechanism proceeded, while three-coordinated Cu (Cu_{3c}) and three-coordinated oxygen (O_{3c}) atoms are terminated on the $CuO/Cu_2O(111)$ surface, which results in a Langmuir-Hinshelwood (LH) mechanism proceeded. These results clearly establish the catalyst structure-catalytic property relationships based on experimental observations and DFT simulations.

Similar results have been extensively reported in many other heterogeneous catalytic reactions, which were used by the calculated E_a values derived from the Arrhenius plots to compare the intrinsic activity [13–15]. Another example is in the identification of the most active Cu facet for low-temperature water gas shift reaction (**Figure 2**) [13]. Three types of Cu nanocrystals were prepared by a morphology-preserved reduction strategy from corresponding Cu_2O nanocrystals. Microscopic structural characterization results and in situ diffuse reflectance infrared Fourier transformed spectra (DRIFTS) of CO adsorption at 123 K confirm the Cu cubes (denoted as c-Cu), octahedra (denoted as o-Cu), and rhombic dodecahedra (denoted as d-Cu), respectively, enclosed with {100}, {111}, and {110} crystal planes (detailed descriptions shown in Ref. [13]). A morphology-dependent catalytic performance was observed on Cu nanocrystal catalyzed low-temperature water gas shift reaction, and the catalytic activity follows an order of c-Cu > d-Cu > d-Cu (**Figure 2A**). However, various Cu nanocrystals are different in particle sizes and surface structures and thus lead to the differently specific BET surface areas and catalytic sites in water gas shift reaction, respectively. As a result, it is difficult to compare the intrinsic activity of c-Cu and d-Cu nanocrystals in

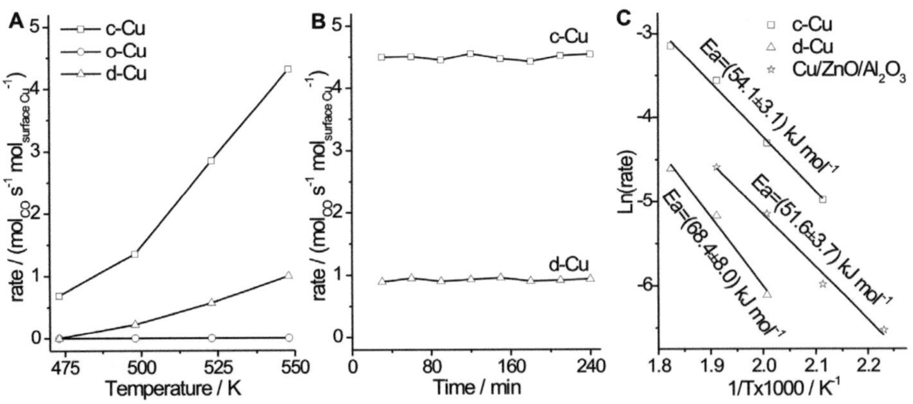

Figure 2.
(A) Reaction rate ($mol_{CO}s^{-1}$ $mol_{surface\ Cu}^{-1}$) of Cu cubes, Cu octahedra, and Cu rhombic dodecahedra in the water gas shift reaction as a function of reaction temperature; (B) reaction rate ($mol_{CO}s^{-1}$ $mol_{surface\ Cu}^{-1}$) of Cu cubes and rhombic dodecahedra at 548 K in the water gas shift reaction as a function of reaction time; (C) the Arrhenius plots of Cu cubes, rhombic dodecahedra, and commercial $Cu/ZnO/Al_2O_3$ catalyst in the water gas shift reaction [13]. Copyright © 2017, Nature Publishing Group.

water-gas shift reaction except o-Cu nanocrystals due to its catalytically inert under the reaction conditions. Subsequently, the Arrhenius equation was plotted based on the catalytic conversions below 15% without the mass transfer effect and heat transfer effect to compare the catalytic performance of intrinsically active sites. The calculated E_a value is 54.1 ± 3.1 kJ mol^{-1} for c-Cu nanocrystals, which is lower than d-Cu nanocrystals with a E_a value of 68.4 ± 8.0 kJ mol^{-1} and similar to commercial $Cu/ZnO/Al_2O_3$ catalyst with a E_a value of 51.6 ± 3.7 kJ mol^{-1} (**Figure 2C**). This indicates that the catalytic reaction proceeds more facile on c-Cu nanocrystals enclosed with {100} crystal planes, and thus, c-Cu nanocrystals are the most active Cu facet for low-temperature water gas shift reaction.

The above-mentioned results are a relatively straightforward and convenient method in the use of the Arrhenius equation. For more in-depth use, the Arrhenius equation can also have been utilized to identify the role and contribution of different active sites of a catalyst nanoparticle to the catalytic reaction based on the calculated E_a and pre-exponential factor A values derived from the Arrhenius equation [8]. c-Cu_2O nanocrystals in different sizes were synthesized by the established methods and subsequently evaluated in the oxidation of CO with excess O_2. The catalytic activities of the restructured CuO/c-Cu_2O catalysts increase with the size decrease (**Figure 3A**), apparently in consistent with their specific BET surface areas. However, their corresponding E_a values are distinctly different. c-Cu_2O nanocrystals in large sizes show a higher E_a value of about 120 kJ mol^{-1} than those in fine sizes with a E_a value of about 80 kJ mol^{-1} (**Figure 3B**), indicating that they exhibit different structures of active sites. Until here, it is still unclear which site contributes to the catalytic performance; thus, the relative relationships of different surface sites to the number of active sites, including facet sites and edge sites, are plotted in order to identify the structure of active sites on different c-Cu_2O nanocrystals. Under the same catalytic reaction conditions, the calculated A values derived from the intercept of the Arrhenius equation are positively related to the density of active sites. Thereby, the calculated ln(A) values of the CuO/c-Cu_2O catalysts were, respectively, plotted as a function of ln(BET surface area) and ln(edge length). As shown in **Figure 3C**, the ln(A) values of CuO/c-Cu_2O-1029, CuO/c-Cu_2O-682, and CuO/c-Cu_2O-446 are proportional to ln(BET surface area) with a slope value of 1.75 ± 0.57, and the ln(A) values of CuO/c-Cu_2O-109 and CuO/c-Cu_2O-34 are proportional to ln(edge length) with a slope value of 0.98. These results are in consistent with the kinetic analysis results of DFT calculations, in which the calculated reaction rate expression based on the elementary reactions exhibits a slope value of 2 between ln(reaction rate) and ln(CuO/Cu_2O(100) surface site) and a slope value of 1 between ln(reaction rate) and ln(CuO/Cu_2O(110) surface site). Furthermore, the reaction orders of CO and O_2 derived from reaction rate equation (described below) in two representative CuO/c-Cu_2O catalysts are calculated (**Figure 3D**), whose values are also in consistent with the DFT calculation results. These kinetic analysis results in combination with DFT calculations successfully demonstrate the structure of active sites with the switch of dominant surface sites contributing to the catalytic activity in CO oxidation reaction from face sites for large c-Cu_2O nanocrystals to edge sites for fine c-Cu_2O nanocrystals.

The Arrhenius equation is also suitable for some complex reactions [16–18], such as the oxidative dehydrogenation of propane (ODHP) to selectively produce propene. You et al. [16] reported that the NbO_x/CeO_2-rod catalysts applied for such a reaction and the calculated E_a values derived from the Arrhenius plots with stable C_3H_8 conversions below 15% without the mass transfer effect and heat transfer effect are dependent on the catalyst structure (**Figure 4**). NbO_x/CeO_2-rod catalysts show a higher E_a value of about 65 kJ mol^{-1} than that of CeO_2 rods with a E_a value of about 45 kJ mol^{-1}, proving that the loading of NbO_x onto CeO_2 rods suppresses their

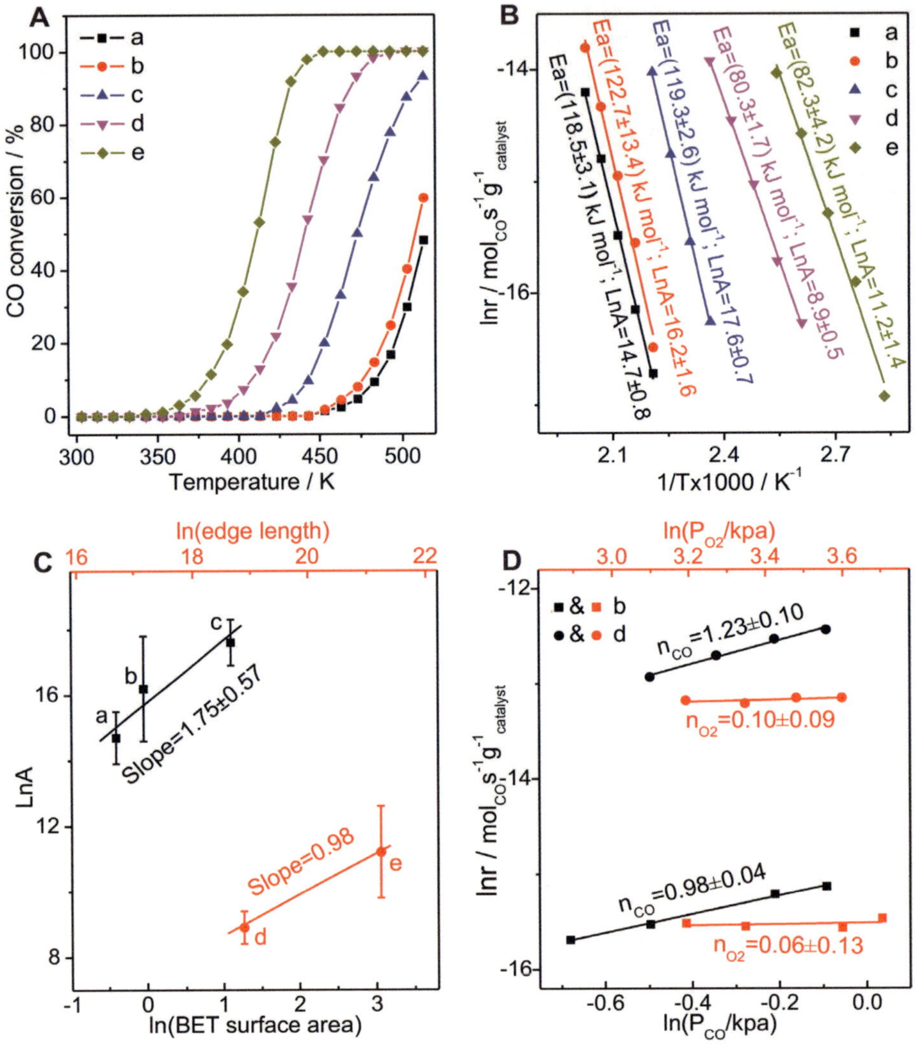

Figure 3.
(A) Stable light-off curves of the CuO/c-Cu_2O structures in different sizes during the oxidation of CO and (B) the corresponding Arrhenius plots; (C) calculated apparent pre-exponential factors (ln A) as a function of the measured BET surface area and the calculated edge length; (D) reaction rates as a function of the partial pressures of CO and O_2: (a) CuO/c-Cu_2O-1029, (b) CuO/c-Cu_2O-682, (c) CuO/c-Cu_2O-446, (d) CuO/c-Cu_2O-109, and (e) CuO/c-Cu_2O-34 [8]. Copyright © 2019, Wiley-VCH Verlag GmbH & Co.

intrinsically catalytic activity. These results clearly demonstrate a wide application for the Arrhenius equation in heterogeneous catalysis. As the science advance, its physical significant is constantly being interpreted with a large amount of analysis of experimental data; therefore, it has a comprehensive guiding significance for chemical kinetics.

2.2 Reaction rate equation and its application in heterogeneous catalysis

Reaction rate equation is an equation of utilizing the concentration or partial pressure of reactants to calculate the reaction rate of chemical reaction. The general expression (Eq. (5)) and differential expression (Eq. (6)) are listed, in which r represents the reaction rate, k represents the rate constant, A and B represent the reactants, and x and y represent the reaction orders of A and B, respectively.

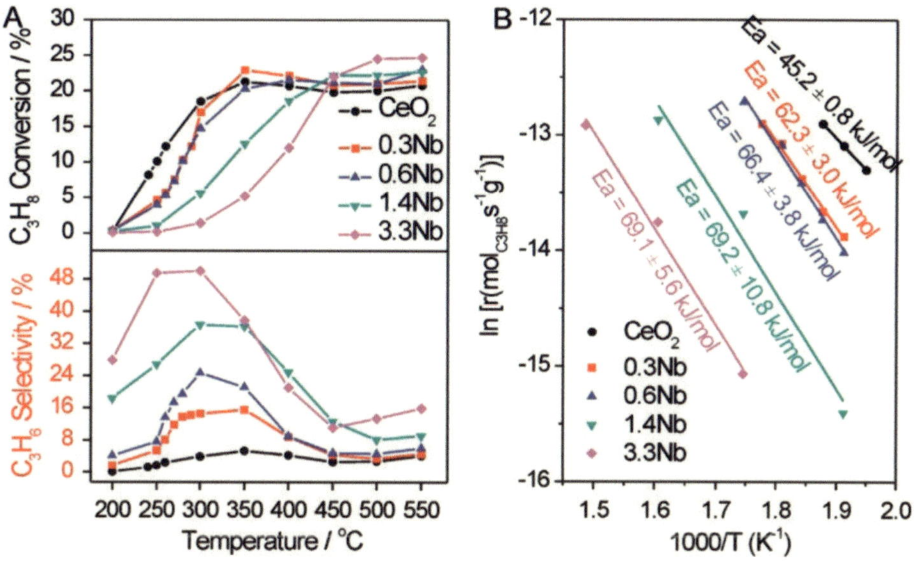

Figure 4.
(A) C_3H_8 conversion and C_3H_6 selectivity of CeO_2 rods and NbO_x/CeO_2 catalysts in the oxidative dehydrogenation of propane reaction and (B) the corresponding Arrhenius plots with calculated apparent activation energies (Ea) [16]. Copyright © 2017, Elsevier.

The reaction orders are determined by the reaction process. Normally, the reaction order is equal to the stoichiometric number of chemical reaction in elementary reaction, while the values are generally not equal in nonelementary reaction. Therefore, it has a complex reaction rate expression in many catalytic reactions that lead to a difficult identification of reaction mechanism.

$$r = k\,[A]^x\,[B]^y \tag{5}$$

$$ln\left(\frac{d[A]}{dt}\right) = lnk + nln[A] \tag{6}$$

For a sample heterogeneous catalytic reaction involving a typical MvK mechanism, such as the above-mentioned CO oxidation reaction in **Figure 3D** [8], the calculated reaction orders of CO and O_2 are 1 and 0, respectively, indicating that the catalytic reaction is only dependent on CO but not on O_2. Such dependence suggests the adsorption of CO and O_2 onto catalyst surface proceeded step by step, as well as the rate-determining step prior to the O_2 adsorption process. Thereby, it can be reasonably proposed that the CO oxidation reaction catalyzed by the facet sites and edge sites of CuO/c-Cu_2O catalysts is both proceeding with the MvK mechanism.

But for complex reactions, the reaction mechanism is hardly proposed by the reaction orders [19–21]. However, the reaction orders are still useful to speculate the important catalytic process, especially in determining the rate-determining step of catalytic reaction [22–25]. A typical example is in the preparation of formic acid, an important chemical in the H_2 storage and other industrial applications, from CO_2 hydrogenation. The support plays a key role in determining the catalytic mechanism of CO_2 hydrogenation into formic acid, which has been confirmed in our previous report by using two common metal oxide (CeO_2 and ZnO) supported Pd catalysts dominantly based on the apparent kinetic analysis and in situ DRIFTS results [22]. The calculated E_a values are similar on Pd/CeO_2 catalysts with different Pd loadings but distinctly different on Pd/ZnO catalysts with different Pd

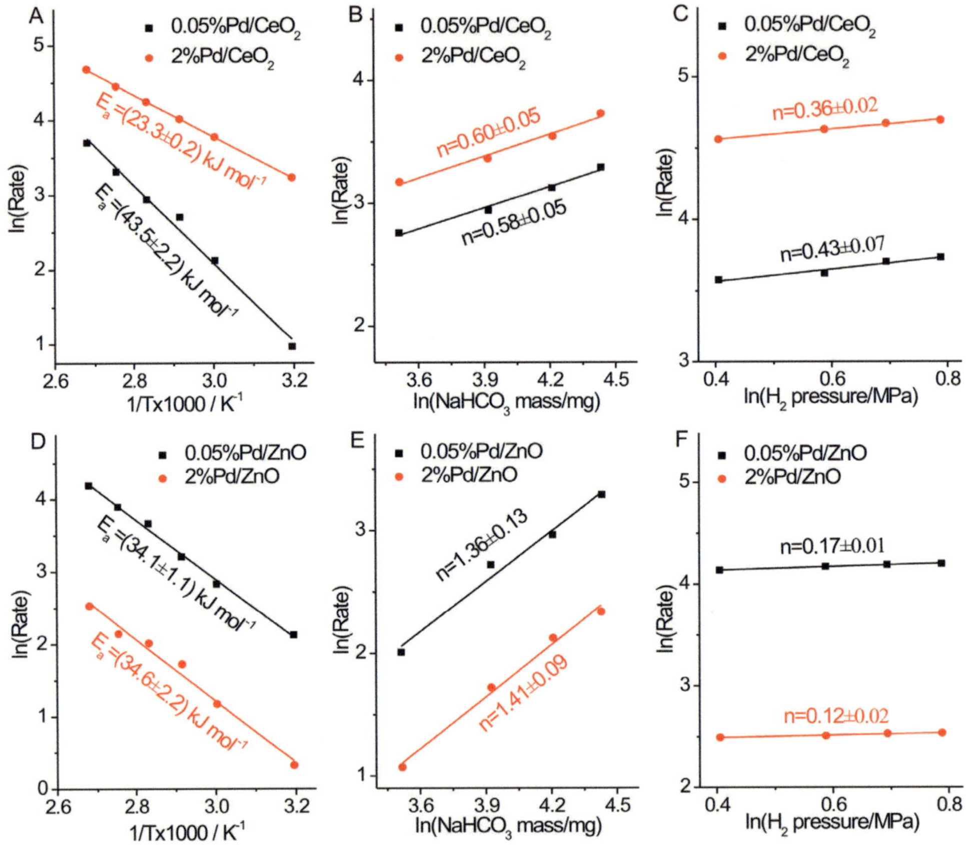

Figure 5.
The Arrhenius plots of (A) 0.05%Pd/CeO_2 and 2%Pd/CeO_2 catalysts and (D) 0.05%Pd/ZnO and 2%Pd/ZnO catalysts; CO_2 reaction orders of (B) 0.05%Pd/CeO_2 and 2%Pd/CeO_2 catalysts and (E) 0.05%Pd/ZnO and 2%Pd/ZnO catalysts at 373 K; and H_2 reaction orders of (C) 0.05%Pd/CeO_2 and 2%Pd/CeO_2 catalysts and (F) 0.05%Pd/ZnO and 2%Pd/ZnO catalysts at 373 K [22]. Copyright © 2019, Elsevier.

loadings (**Figure 5A** and **D**), suggesting that the catalytic performance on Pd/CeO_2 catalysts is affected by the Pd structures. The reaction orders of CO_2 and H_2 clearly demonstrate that the rate-determining step of CO_2 hydrogenation to formic acid on Pd/ZnO catalysts involves a CO_2-contained elementary reaction due to a relatively higher CO_2 reaction order (~1.4) (**Figure 5B** and **E**) and of CO_2 hydrogenation to formic acid on Pd/CeO_2 catalysts involves a H_2-contained elementary reaction due to a relatively higher H_2 reaction order (~0.4) (**Figure 5C** and **F**). H_2 is generally activated on Pd, and thus, the structure of Pd affects the catalytic performance of Pd/CeO_2 catalysts but not that of Pd/ZnO catalysts, in consistent with their calculated E_a results.

This conclusion is further proved by Pd/ZrO_2 catalyzed CO_2 hydrogenation into formate (**Figure 6**) [23]. Experimental observation results in CO_2-TPD profiles (detailed description shown in Ref. [23]) showed the basicity densities following an order of 2%Pd/ZrO_2-T (Tetragonal ZrO_2) > 2%Pd/ZrO_2-M&T (Mixed ZrO_2) ≈ 2%Pd/ZrO_2-M (Monoclinic ZrO_2), in consistent with the intrinsic activity order. Comparing the reaction orders of H_2 and CO_2 on 2%Pd/ZrO_2-T and 2%Pd/ZrO_2-M catalysts, similar H_2 reaction orders were observed, while the CO_2 reaction orders were higher on 2%Pd/ZrO_2-M catalyst than that on 2%Pd/ZrO_2-T catalyst. These results clearly demonstrate that the catalytic performance of Pd/ZrO_2 catalyst in CO_2 hydrogenation into formate is strongly dependent on the

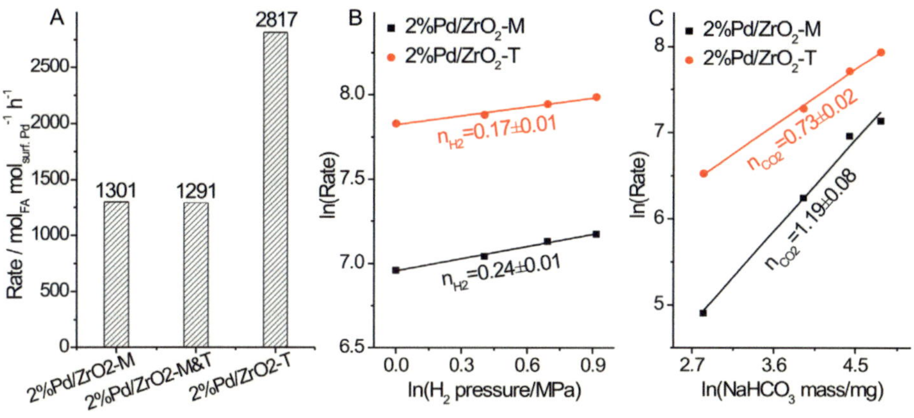

Figure 6.
(A) Production rate (mol_{HCOO^-} $mol_{surf.\ Pd}^{-1}$ h^{-1}) of formate on representative various 2%Pd/ZrO_2 catalysts based on the surface Pd atoms; reaction orders of (B) H_2 and (C) bicarbonate of 2%Pd/ZrO_2-M and 2%Pd/ZrO_2-T catalysts [23]. Copyright © 2019, Elsevier.

surface basicity densities, 2%Pd/ZrO_2-T support holds the higher surface basicity densities, and thus, Pd/ZrO_2-T catalyst is more active in catalyzing the formate production.

Reaction rate equation is the real reflection of chemical reaction rate. It has been widely used as an important method in investigating the catalytic mechanism. However, several factors, including temperature, chemical reaction, concentration, pressure, order, solvent, light, and catalyst, strongly affect the rate of a chemical reaction. Among them, temperature usually plays a key factor. Generally, the rate of a chemical reaction increases with an increase in reaction temperature because higher kinetic energy can result in more collisions between the reactant molecules. The increase of collision chance will cause the catalytic reaction more likely to happen. The temperature effect on reaction rate can be quantified by the Arrhenius equation. Notably, some reaction rates are negatively affected by the temperature, while some others are independent of temperature. The chemical reaction, concentration, pressure, and order directly determine the reaction rate. In chemical reaction, the complexity of the reaction and the state of the reactants are important. Such as, the reaction of a powder in a solution always proceeds faster than the reaction of a large chunk of a solid. Concentration, pressure, and order are the important parameters of reaction rate equation and thus directly affect the reaction rate. Solvent, light, and catalyst are the external factors that generally do not participate in a reaction but affect the reaction rate. In familiar with these influence factors, it can guide us to better understand and apply the reaction rate equation.

3. Summary and outlook

In summary, this chapter introduced the Arrhenius equation and reaction rate equation, two important equations in chemical kinetics, and their applications in heterogeneous catalysis in detail. By the analysis of some specific reactions in the documents, the Arrhenius equation could be used to calculate the E_a and A values in catalytic reaction, which were related to the intrinsic activity of the number of active sites, respectively. The reaction rate equation could be used to calculate the reaction order of a single reactant that was useful in speculating the contribution of a single reactant in catalytic reaction. Through these investigations of chemical kinetic equations in catalytic reaction and further combining with DFT calculations, it well proposed the structure of active sites and the catalytic mechanism in catalytic reaction and thus established the concept of catalyst structure-catalytic property relationships. These would be promising to guide the structural design and controlled synthesis of novel efficient catalytic materials in the future.

Nowadays, the development of chemical kinetics is very rapid and the contained fields gradually enlarged with the research mode from basement state turning to excitation state, from small molecule turning to big molecule, and from gas phase turning to condensed phase. Benefiting from the improvement of advanced characterization skills, especially in the development of surface analysis and intermediate tracking skills, the catalytic science has become the frontier field of chemical kinetics. Furthermore, the progress of basic theory method also accelerates the application and development of chemical kinetics in catalytic science. The combination of theory and experiment can help to gain a deep insight into how various elementary reactions occur. The combination of more precise experimental results with more precise theoretical research is a powerful driving force for the continued development of this research field. Practice once again proves that the close combination of theory and experiment is the only way for the development of science.

Acknowledgements

Thanks to Li-Ping Fan and Yue-Juan Wang who provided the partial writing materials and also commented on the manuscript. This work was financially supported by Shuang-long Scholars Program, Special Foundation of Youth Scholars of Zhejiang Normal University, and Natural Science Foundation of Zhejiang Province (LQ20B030007).

Conflict of interest

The authors declare no conflict of interest.

Author details

Zhenhua Zhang*, Li-Ping Fan and Yue-Juan Wang
Key Laboratory of the Ministry of Education for Advanced Catalysis Materials, Institute of Physical Chemistry, Zhejiang Normal University, Jinhua, China

*Address all correspondence to: hanyuzh@mail.ustc.edu.cn

IntechOpen

© 2020 The Author(s). Licensee IntechOpen. Distributed under the terms of the Creative Commons Attribution - NonCommercial 4.0 License (https://creativecommons.org/licenses/by-nc/4.0/), which permits use, distribution and reproduction for non-commercial purposes, provided the original is properly cited. (cc) BY-NC

References

[1] Gillespie DT. Deterministic limit of stochastic chemical kinetics. Journal of Physical Chemistry B. 2009;**113**: 1640-1644. DOI: 10.1021/jp806431b

[2] Gillespie DT. Stochastic simulation of chemical kinetics. Annual Review of Physical Chemistry. 2007;**58**:35-55. DOI: 10.1146/annurev.physchem.58.032806.104637

[3] Frennet A, Hubert C. Transient kinetics in heterogeneous catalysis by metals. Journal of Molecular Catalysis A: Chemical. 2000;**163**:163-188. DOI: 10.1016/S1381-1169(00)00385-X

[4] Huang W. Crystal plane-dependent surface reactivity and catalytic property of oxide catalysts studied with oxide nanocrystal model catalysts. Topic in Catalysis. 2013;**56**:1363-1376. DOI: 10.1007/s11244-013-0139-6

[5] Huang W. Surface chemistry of solid catalysts. Scientia Sinica Chimica. 2018;**48**:1076-1093. DOI: 10.1360/N032018-00033

[6] Chen S, Xiong F, Huang W. Surface chemistry and catalysis of oxide model catalysts from single crystals to nanocrystals. Surface Science Reports. 2019;**74**:100471. DOI: 10.1016/j.surfrep.2019.100471

[7] Bao H, Zhang W, Hua Q, Jiang Z, Yang J, Huang W. Crystal-plane-controlled surface restructuring and catalytic performance of oxide nanocrystals. Angewandte Chemie International Edition. 2011;**50**:12294-12298. DOI: 10.1002/anie.201103698

[8] Zhang Z, Wu H, Yu Z, Song R, Qian K, Chen X, et al. Site-resolved Cu_2O catalysis in the oxidation of CO. Angewandte Chemie International Edition. 2019;**58**:4276-4280. DOI: 10.1002/anie.201814258

[9] Bai Y, Zhang W, Zhang Z, Zhou J, Wang X, Wang C, et al. Controllably interfacing with metal: A strategy for enhancing CO oxidation on oxide catalysts by surface polarization. Journal of the American Chemical Society. 2014;**136**:14650-14653. DOI: 10.1021/ja506269y

[10] Bao H, Zhang Z, Hua Q, Huang W. Compositions, structures, and catalytic activities of CeO_2@Cu_2O nanocomposites prepared by the template-assisted method. Langmuir. 2014;**30**:6427-6436. DOI: 10.1021/la501406w

[11] Zhang Z, Song R, Yu Z, Huang W. Crystal-plane effect of Cu_2O templates on compositions, structures and catalytic performance of Ag/Cu_2O nanocomposites. CrystEngComm. 2019;**21**:2002-2008. DOI: 10.1039/c8ce02164c

[12] Li D, You R, Yang M, Liu Y, Qian K, Chen S, et al. Morphology-dependent evolutions of sizes, structures, and catalytic activity of Au nanoparticles on anatase TiO_2 nanocrystals. Journal of Physical Chemistry C. 2019;**123**:10367-10376. DOI: 10.1021/acs.jpcc.9b00262

[13] Zhang Z, Wang S-S, Song R, Cao T, Luo L, Chen X, et al. The most active Cu facet for low-temperature water gas shift reaction. Nature Communications. 2017;**8**:488-497. DOI: 10.1038/s41467-017-00620-6

[14] Bu Y, Weststrate CJ, Niemantsverdriet JW, Fredriksson HOA. Role of ZnO and CeO_x in Cu-based model catalysts in activation of H_2O and CO_2 dynamics studied by in situ ultraviolet-visible and X-ray photoelectron spectroscopy. ACS Catalysis. 2016;**6**:7994-8003. DOI: 10.1021/acscatal.6b02242

[15] Graciani J, Mudiyanselage K, Xu F, Baber AE, Evans J, Senanayake SD,

et al. Highly active copper-ceria and copper-ceria-titania catalysts for methanol synthesis from CO_2. Science. 2014;**345**:546-550. DOI: 10.1126/science.1253057

[16] You R, Zhang X, Luo L, Pan Y, Pan H, Yang J, et al. NbO_x/CeO_2-rods catalysts for oxidative dehydrogenation of propane: Nb-CeO_2 interaction and reaction mechanism. Journal of Catalysis. 2017;**348**:189-199. DOI: 10.1016/j.jcat.2016.12.012

[17] Sun B, Gu C, Ma J, Liang B. Kinetic study on TEMPO-mediated selective oxidation of regenerated cellulose. Cellulose. 2005;**12**:59-66. DOI: 10.1023/B:CELL.0000049409.56806.da

[18] Sinev I, Kardash T, Kramareva N, Sinev M, Tkachenko O, Kucherov A, et al. Interaction of vanadium containing catalysts with microwaves and their activation in oxidative dehydrogenation of ethane. Catalysis Today. 2009;**141**:300-305. DOI: 10.1016/j.cattod.2008.04.021

[19] Jia A-P, Hu G-S, Meng L, Xie Y-L, Lu J-Q, Luo M-F. CO oxidation over CuO/$Ce_{1-x}Cu_xO_{2-\delta}$ and $Ce_{1-x}Cu_xO_{2-\delta}$ catalysts: Synergetic effects and kinetic study. Journal of Catalysis. 2012;**289**:199-209. DOI: 10.1016/j.jcat.2012.02.010

[20] Li N, Chen Q-Y, Luo L-F, Huang W-X, Luo M-F, Hu G-S, et al. Kinetic study and the effect of particle size on low temperature CO oxidation over Pt/TiO_2 catalysts. Applied Catalysis B: Environmental. 2013;**142-143**:523-532. DOI: 10.1016/j.apcatb.2013.05.068

[21] Liu H-H, Wang Y, Jia A-P, Wang S-Y, Luo M-F, Lu J-Q. Oxygen vacancy promoted CO oxidation over Pt/CeO_2 catalysts: A reaction at Pt–CeO_2 interface. Applied Surface Science. 2014:314725-314734. DOI: 10.1016/j.apsusc.2014.06.196

[22] Zhang Z, Zhang L, Yao S, Song X, Huang W, Hülsey MJ, et al. Support-dependent rate-determining step of CO_2 hydrogenation to formic acid on metal oxide supported Pd catalysts. Journal of Catalysis. 2019;**376**:57-67. DOI: 10.1016/j.jcat.2019.06.048

[23] Zhang Z, Zhang L, Hülsey MJ, Yan N. Zirconia phase effect in Pd/ZrO_2 catalyzed CO_2 hydrogenation into formate. Molecular Catalysis. 2019;**475**:110461. DOI: 10.1016/j.mcat.2019.110461

[24] Harrington DA. The rate-determining step in electrochemical impedance spectroscopy. Journal of Electroanalytical Chemistry. 2015;**737**:30-36. DOI: 10.1016/j.jelechem.2014.06.003

[25] Oh H-S, Nong H-N, Reier T, Gliech M, Strasser P. Oxide-supported Ir nanodendrites with high activity and durability for the oxygen evolution reaction in acid PEM water electrolyzers. Chemical Science. 2015;**6**:3321-3328. DOI: 10.1039/c5sc00518c

Section 3

Fenton and Ferrate in Wastewater Treatment

Chapter 4

Advancements in the Fenton Process for Wastewater Treatment

Min Xu, Changyong Wu and Yuexi Zhou

Abstract

Fenton is considered to be one of the most effective advanced treatment processes in the removal of many hazardous organic pollutants from refractory/toxic wastewater. It has many advantages, but drawbacks are significant such as a strong acid environment, the cost of reagents consumption, and the large production of ferric sludge, which limits Fenton's further application. The development of Fenton applications is mainly achieved by improving oxidation efficiency and reducing sludge production. This chapter presents a review on fundamentals and applications of conventional Fenton, leading advanced technologies in the Fenton process, and reuse methods of iron containing sludge to synthetic and real wastewaters are discussed. Finally, future trends and some guidelines for Fenton processes are given.

Keywords: Fenton, Fenton-like, Fenton sludge, reuse, application

1. Introduction

The presence of many organic pollutants in wastewater, surface water, and groundwater may result from contaminated soil, agricultural runoff, industrial wastewater, and hazardous compounds storage leakage [1]. These organic pollutants, such as volatile phenols, benzene, and benzene derivatives, are considered highly toxic and low biodegradable. In some instances, conventional treatment methods such as biological processes are not sufficient to remove them. In order to improve water quality, advanced treatment is needed to remove the refractory organics.

Fenton is an effective advanced treatment process. The hydroxyl radical (·OH) can be generated from the reaction between aqueous ferrous ions and hydrogen peroxide (H_2O_2), and it can destroy refractory and toxic organic pollutants in wastewater [2]. Fenton discovered the Fenton reaction in 1894, and he reported that H_2O_2 could be activated by ferrous (Fe^{2+}) salts to oxidize tartaric acid. However, its application as an oxidizing process for destroying toxic organics was achieved until the late 1960s [3]. Fenton was mainly used to treat wastewater by radical oxidation and flocculation. H_2O_2 is catalyzed by ferrous ions to decompose into HO· and to trigger the production of other radicals, which can fully oxidize organic matters. The hydroxyl radical (·OH) has a strong oxidation capacity (standard potential = 2.80 V versus standard hydrogen electrode) [4]. Hydroxyl radicals can effectively oxidize refractory organic pollutants in wastewater and even completely mineralized them into CO_2, water, and inorganic salts [5]. Meanwhile,

IntechOpen

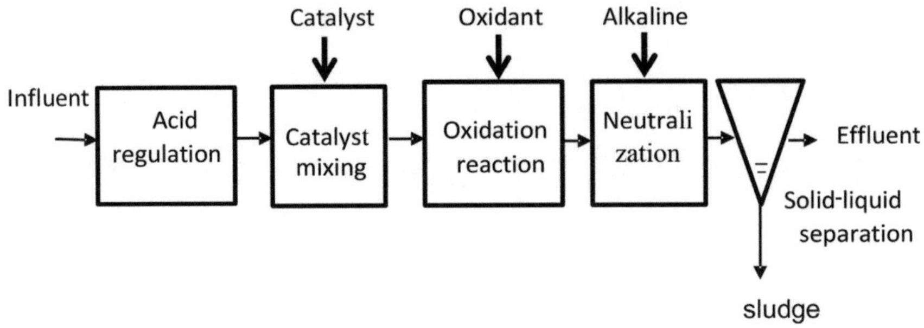

Figure 1.
Schematic diagram of the Fenton process.

the iron complex produced in the treatment of wastewater by Fenton will play the role of flocculant.

The conventional Fenton continuous flow process configuration, as illustrated in **Figure 1**, including acid regulation, catalyst mixing, oxidation reaction, neutralization, and solid–liquid separation. Fenton has many advantages, such as its high performance and simplicity (operated at room temperature and atmospheric pressure) for the oxidation of organics [6] and its non-toxicity [7] (H_2O_2 can break down into environmentally safe species like H_2O and O_2). However, Fenton also has some inherent disadvantages, which limit its application and promotion. For example, strict pH range, high H_2O_2 consumption, and the accumulation of ferric sludge that affects the oxidation efficiency [8, 9]. In order to overcome these disadvantages, the enhancement of the Fenton process has attracted much attention from researchers. Both heterogeneous and homogeneous catalysts were used to replace Fe^{2+}, including ferric oxide [10], iron minerals [11], and nano zero-valent iron [12].

On the one hand, the loss of iron and the consequent sludge generation can be reduced by using reductant or transition metal, giving rise to a heterogeneous Fenton process. On the other hand, some external energy was introduced into the Fenton and Fenton-like processes to form photo-Fenton/Fenton-like processes [13, 14], electro-Fenton-like processes [15, 16], and so on [17, 18]. Hence, this work mainly summarizes the recent advancements in the Fenton process related to improving Fenton oxidation efficiency and minimizing sludge production. It also describes the main drawbacks and potential applications based on recent developments. Some recommendations are also stated in the Conclusions section.

2. Fenton process

Currently, two mechanisms have been proposed to explain the degradation of organic matters by Fenton reaction. One is the Harber-Weiss mechanism [19], which considered that active oxide species ·OH are generated to degrade organics in Fenton reaction. The other is the mechanism of high iron oxide intermediates, which was proposed by Bray and Gorin [20]. They suggested that the strong oxidizing iron substances (FeO^{2+}, FeO^{3+}) were produced in Fenton reaction, rather than ·OH. With the development of spectroscopy and chemical probe method, it is generally accepted that the formation of ·OH initiates the Fenton oxidation.

The traditionally accepted Fenton mechanism is represented by Eqs. (1)–(7) [21].

$$Fe^{2+} + H_2O_2 \rightarrow Fe^{3+} + \cdot OH + OH^- \qquad k_1 = 63.5\,M^{-1} \cdot s^{-1} \qquad (1)$$

$$Fe^{2+} + \cdot OH \rightarrow Fe^{3+} + OH^- \quad k_2 = 3.2 \times 108\ M^{-1} \cdot s^{-1} \tag{2}$$

$$RH + \cdot OH \rightarrow H_2O + R \rightarrow \text{further oxidation} \tag{3}$$

$$R \cdot + Fe^{3+} \rightarrow Fe^{2+} + R^+ \tag{4}$$

$$Fe^{3+} + H_2O_2 \rightarrow Fe^{2+} + \cdot OOH + H^+ \quad k_2 = 0.001\text{–}0.01\ M^{-1} \cdot s^{-1} \tag{5}$$

$$\cdot OH + H_2O_2 \rightarrow \cdot OOH + H_2O \quad k_2 = 3.3 \times 107\ M^{-1} \cdot s^{-1} \tag{6}$$

$$\cdot OH + \cdot OH \rightarrow \cdot OOH + OH^- \quad k_2 = 6.0 \times 109\ M^{-1} \cdot s^{-1} \tag{7}$$

According to the above equation, ferrous iron (Fe^{2+}) was rapidly oxidized to ferric ions (Fe^{3+}), while Fe^{2+} is regenerated from the so-called Fenton-like reaction between Fe^{3+} and H_2O_2 at a very slow rate. Equation (1) is usually considered as the core of the Fenton reaction.

The Fenton process is usually operated under the solution pH value of 3. The oxidation activity of ·OH is related to the solution pH. The oxidation potential of ·OH increases and the oxidation capacity is enhanced with decreasing pH [22, 23]. In addition, the activity of Fenton reagent is reduced with increasing pH due to the lack of active Fe^{2+}, in which the formation of inactive iron oxohydroxides and ferric hydroxide precipitate. Meanwhile, auto-decomposition of H_2O_2 appears at high pH [24].

At very low pH values, iron complex species $[Fe(H_2O)_6]^{2+}$ and stable oxonium ion $[H_3O_2]^+$ exist, which reduces the reactivity between Fe^{2+} and H_2O_2 [25, 26]. Therefore, the efficiency of the Fenton process to degrade organic compounds is reduced both at high and low pH. In addition, there are many competitive reactions in the Fenton reaction system. In Fenton oxidation, the reaction rate is dependent on the iron dosage, while the extent of mineralization is directly related to the concentration of oxidant.

It is important to understand the mutual relationships between Fenton reagent in terms of ·OH production and consumption. These relationships were investigated and classified them into three categories according to the quantity of the $[Fe^{2+}]_0/[H_2O_2]_0$ ratio (initial concentration of Fe^{2+} versus initial concentration of H_2O_2) [27]. Their results showed that $[Fe^{2+}]_0/[H_2O_2]_0$ ratio and organics can affect the competition in Fenton reaction paths.

The main disadvantages for the application of the Fenton process are the relatively high cost of H_2O_2 and the high amount of ferric sludge produced in the neutralization step of the treated solution before disposal. These drawbacks and the more increasingly stringent water regulations are a challenge to develop solutions addressed to improve the Fenton technology. On one hand, the energy was introduced into Fenton to enhance the ·OH generation, such as photo-Fenton, electro-Fenton, and so on. On the other hand, iron-based catalysts and reuse of Fenton sludge were developed as a Fenton-like reaction.

3. Enhanced Fenton process

3.1 Photo-Fenton process

The classic Fenton reaction efficiency was affected by the conversion rate from Fe^{3+} to Fe^{2+}. Recent methods have promoted the in situ circulation from Fe^{3+} to Fe^2. A combination of hydrogen peroxide and UV radiation with Fe^{2+} or Fe^{3+} oxalate ion [photo-Fenton (PF) process] produces more ·OH compared to the conventional Fenton method [28]. The hydroxy-Fe^{3+} complexes after Fenton reaction mainly exists in the form of $Fe(OH)^{2+}$ at pH 2.8–3.5. The photochemical regeneration of Fe^{2+} by photo-reduction (Eq. (8)) of ferric Fe^{3+} occurs in the photo-Fenton reaction. The newly generated Fe^{2+} reacts with H_2O_2 and generates $\cdot OH$ and Fe^{3+}, and the cycle continues:

$$Fe(OH)^{2+} + hv \rightarrow Fe^{2+} + \cdot OH \tag{8}$$

Direct photolysis of H_2O_2 (Eq. (9)) produces $\cdot OH$, which can be used for the degradation of organic compounds, and in turn increases the rate of degradation of organic pollutants [29].

$$H_2O_2 + hv \rightarrow 2 \cdot OH \tag{9}$$

However, photo-Fenton gives a better degradation of low concentration organic pollutants. Because the high concentration organic pollutants could reduce the absorb radiation of iron complex, which needs a longer radiation time and more H_2O_2 dosage.

Excess H_2O_2 can easily capture $\cdot OH$. In order to improve the efficiency of photo-Fenton, several organic ligands such as EDTA, EDDS, oxalate, and other organic carboxylic acid were added and complexed with Fe^{3+} under photocatalysis [29, 30]. The positive effects achieved by these ions can be attributed to the following aspects: (i) iron-ligands having higher ability compete for UV light in a wide wavelength range compared to other Fe^{3+}-complexes, and promoting the reduction of ferric ion to ferrous ion and accordingly, regeneration of higher amounts of $\cdot OH$, (ii) Promoting H_2O_2 activation and $\cdot OH$ radical generation, (iii) improving iron dissolution at pH 7.0, and (iv) operating over the broad range of the solar radiation spectrum [14].

Compared with the classic Fenton, photo-Fenton has many advantages. A photo-induced Fe^{3+}/Fe^{2+} redox cycle could decrease the dosage of catalyst in Fenton, which effectively reduce the formation of iron sludge [31–33]. Meanwhile, solar or UV light can increase the utilization of H_2O_2, and possess photolysis on several small molecule organics. However, photo-Fenton has many disadvantages, such as low utilization of visible light, the required UV energy for a long time, high energy consumption, and cost.

3.2 Electro-Fenton process

Electrochemical processes can be combined with Fenton processes (EF processes) during WW treatment to improve the Fenton processes. Fe^{2+} and H_2O_2 were produced by the electrochemical method as Fenton reagent. The electro-Fenton process follows the reaction shown below [34, 35], where H_2O_2 can be generated in situ via a two-electron reduction of dissolved oxygen on the surface of the cathode in an acidic solution when the electrochemical process is applied (reaction Eq. (10)).

$$O_2 + 2H^+ + 2e^- \rightarrow H_2O_2 \tag{10}$$

Also, the produced ferric ion from Eq. (1) can be reduced to Fe^{2+} by electrochemical regeneration of Fe^{3+} ions on the cathode surface:

$$Fe^{3+} + e^- \rightarrow Fe^{2+} \tag{11}$$

Water was oxidized to oxygen ta the anode (Eq. (12)):

$$2H_2O \rightarrow O_2 + 4H^+ + 4e^- \tag{12}$$

·OH was also generated at the surface of a high-oxygen overvoltage anode from water oxidation:

$$H_2O + H_2O_2 \rightarrow H^- + \cdot OH + e^- \tag{13}$$

Compared with tradition Fenton reaction, electro-Fenton has certain advantages, including (i) the production of H_2O_2 in situ via an electrochemical processes is beneficial for an increase in the organics degradation efficiency, a decrease in the cost, and a reduction in the risks associated with transportation; (ii) ferrous ion is regenerated through the reduction of ferric ions on the cathode, which reduces the production of iron sludge; and (iii) realizing the diversification of organics degradation pathway, such as Fenton oxidation, anodic oxidation, flocculation, and electric adsorption [36].

Electro-Fenton gives a better degradation of alachlor than the tradition Fenton. However, electro-Fenton processes have some problems with respect to H_2O_2 production. The production of H_2O_2 is slow because oxygen has low solubility in water and the current efficiency under reduced pH ($pH < 3$) is low. In addition, the efficiency of the electro-Fenton process depends on electrode nature, pH, catalyst concentration, electrolytes, dissolved oxygen level, current density, and temperature [37].

3.3 Sono-Fenton process

The combined treatment using ultrasound with Fenton reagent is known as sono-Fenton, which provides a synergistic effect on organic degradation [38, 39]. Ultrasound can enhance the Fenton's oxidation rate due to the generation of more ·OH caused by the cavitation within ultrasonic irradiation.

The physical effect of cavitation is the generation of intense convection in the medium through the phenomena of microturbulence and shock waves, whereas the chemical effect of cavitation is the generation of radical species, such as oxygen (·O), hydroperoxyl (·OOH), and ·OH through the dissociation of solvent vapor during transient collapse of cavitation bubbles [40]. On the other hand, Fe^{3+} continuously reacts with H_2O_2, according to Eq. (5). A part of Fe^{3+} after a Fenton reaction exists in the form of $Fe{=}OOH^2$. $Fe{=}OOH^2$ can be quickly decomposed into Fe^{2+} and ·OOH; thereby, the phenomenon promotes the Fe^{3+}/Fe^{2+} redox cycle [41]. In addition, ultrasound provides stirring and mass transfer effects to promote the diffusion of reactants in solution and improve the efficiency of Fenton reaction. However, sono-Fenton has some disadvantages, such as high cost and energy-intensive, so it is limited in practical application.

3.4 Fenton combined with other wastewater treatment technologies

The physical, chemical, and biological technologies have been widely used to treat wastewater. However, most of the real wastewater contains many organic pollutants with high toxicity and low biodegradability [42]. So that biological technologies are not enough. Moreover, the physical and chemical techniques are often effective for color, macromolecular organics, and suspended matter removal.

In order to improve the wastewater quality, advanced treatment needs the removal of refractory organics from the wastewater. In recent years, Fenton has been used to treat refractory wastewater by combining it with other wastewater-treatment technologies. The application of the Fenton oxidation process, as a pretreatment, oxidizes refractory organics and improves the biodegradability, solubility, and coagulation, which are beneficial to the subsequent treatment.

Fenton was proved to be a feasible technique as the pre-oxidation for polluted pharmaceutical wastewater. Fenton as a pre-treatment process increased the BOD_5/COD value from 0.26 to 0.5, and also removed chloramphenicol, diclofenac, p-aminophenol, benzoic acid, and other toxic organics in the final effluent results from the Fenton-biological treatment processes [43].

Compared with biological treatment alone, Fenton-biological treatment processes improved COD removal [44]. Fenton combined with membrane filtration was also used to treat pharmaceutical wastewater. Although single nanofiltration (NF) and Fenton could effectively remove pharmaceutical active compounds, high organic load favored membrane fouling, and resulting in flux decline. The calcium salts were found to be the main fouling on the NF membrane surface. Fenton as the pre-treatment could ensure higher flux for the NF process. Also, the Fenton-NF system was found to be a promising method for wastewater from the pharmaceutical industry containing etodolac [45].

On the other hand, lime/unhair effluent, which contain highly loaded organic hazardous wastes, have been effectively treated by the Fenton-membrane filtration system at the pilot-scale [46]. Fenton was used also as a post-treatment of other treatment technologies for degrading residual organics, including physico-chemical and biological treatment. Compared with the aerobic sequencing batch reactor (SBR), SBR-Fenton could improve the TOC removal rate of textile wastewater up to 12% [47]. The solar photo-Fenton process was found to be an efficient process in removing phytotoxicity from Olive mill wastewater samples [48]. A COD and phenol removal as high as 94 and 99.8% could be achieved in coagulation/flocculation combined with the solar photo-Fenton system [48]. This combined technology, taking into account its reasonable overall cost, can be applied in somewhere with plenty of sunshine. Based on the technological and economic analysis performed, all combined treatment technologies will provide better performance than single treatment.

The Fenton process, combined with biological technologies, has shown quite low operational cost. The integrated Fenton-membrane processes were found efficient in removing organics from the industry wastewater. The phenolic compounds concentrated in the concentrate flow by membrane filtration could be recovered and further valorized in various industries. Although the combined Fenton and membrane technologies show the relatively higher overall cost of the combined membrane technologies, it is necessary to estimate the accuracy of the potential profit from the sale and valorization of these by-products recovered by this process.

4. The application of modified iron source as heterogeneous catalysts in Fenton reactions

The Fenton reaction in which iron salts are used as a catalyst is defined as a homogeneous Fenton process. Nevertheless, there are some disadvantages, including (i) the formation of ferric hydroxide sludge at pH values above 4.0 and its removal, (ii) difficulty in catalyst recycle and reuse, (iii) high energy consumption, and (iv) limitation of operating pH range. Therefore, the application of modified iron source as heterogeneous catalysts in Fenton reaction to overcome the shortcomings of homogeneous catalysis has been widely studied. Different heterogeneous catalysts have been used in Fenton reactions, including zero valent iron [49], iron pillared clays [50], and iron minerals [51]. **Figure 2** shows various types of heterogeneous Fenton-like catalysts.

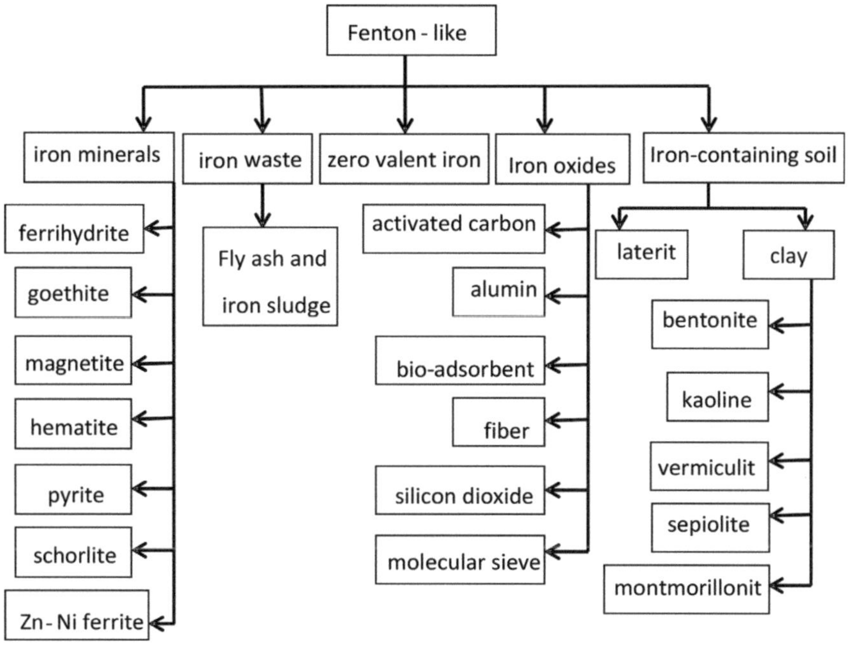

Figure 2.
Iron-containing catalysts.

4.1 Fenton-like reactions using zero-valent iron

Recently, zero-valent iron (Fe^0) has been increasingly used in the heterogeneous Fenton system, due to its large specific surface area and high reactivity. It is reported that the removal of contaminants in the Fe^0 induced heterogeneous Fenton system involves two steps [52–54]: (i) H_2O_2 decomposes on or near the Fe^0 surface to form Fe^{2+} (Eq. (14)); (ii) then, Fenton reaction occurs, Fe^{2+} reacts with H_2O_2 to produce ·OH (Eq. (1)), and contaminants are degraded. Meanwhile, the produced Fe^{3+} is further reduced to Fe^{2+} (Eq. (15)).

$$Fe^0 + H_2O_2 + 2H^+ \rightarrow Fe^{2+} + 2H_2O \tag{14}$$

$$Fe^{3+} + Fe^0 \rightarrow 3Fe^{2+} \tag{15}$$

The degradation of trichloroethylene (TCE) in nano-scale zero-valent iron (nZVI) Fenton systems with Cu(II) was investigated [55]. TCE was significantly degraded (95%) in 10 min in the nZVI Fenton system with 20 mM Cu(II) at initial pH 3, while slight degradation (25%) was observed in nZVI Fenton system without Cu(II) at the same experimental condition. Because of the high activity on the Fe^0 surface, Fe^0 could easily coalesce into aggregates, which reduced the reactivity. The particle size of Fe^0 is too small to recycle and separate at the end of the treatment.

4.2 Fenton-like reactions using iron oxides

Different physicochemical characteristics of these oxides make them favorable for oxidative reactions, where the surface area, pore size/volume, and the crystalline structure have significant effects on their activities. The amorphous Fe_2O with the largest surface area has the lowest catalytic efficiency, while the higher efficiency was achieved with crystalline α-Fe_2O_3, which has a significantly lower surface area [56]. In the Fenton process, magnetite has gained considerable attention than other iron oxides due to its unique characteristic: the magnetically easy separation of magnetite catalysts from the reaction system as a result of its magnetic property.

In many cases, magnetite offered better performance due to the presence of Fe^{2+} cations in its structure [57]. A comparison of the catalytic activity of amorphous iron (III) oxide, maghemite, magnetite, and goethite mixed with quartz was carried out by Hanna et al. [58] for methyl red degradation in presence of H_2O_2. The authors indicated that the oxidation state of iron in the oxides was the critical parameter, considering that Fe^{2+} is superior to Fe^{3+} in Fenton processes. In this study, magnetite exhibited the highest rate constant normalized to surface area per unit mass of oxide (SSA) at neutral pH value.

4.3 Fenton-like reactions using other iron-containing catalysts

The characteristic of iron materials mainly determines the oxidation efficiency in the heterogeneous Fenton reaction. The design of iron-containing catalysts can improve the activity and stability of the heterogeneous Fenton reaction. In recent years, heterogeneous Fenton catalysts have also been developed in the following aspects: (1) iron-loaded material, that is, iron is loaded on the porous materials, such as carbon nanotubes, clay and molecular sieves by a simple method. These materials are considered as potential heterogeneous Fenton catalysts because of low cost, high specific surface area, rich active sites, and easy separation, (2) new iron-containing materials. In the homogeneous Fenton-like processes, the catalysts used in the Fenton-like processes include Fe^{3+}, Cu^{2+}, Mn^{2+}, Co^{2+}, and Ag^{+}. Sometimes organic or inorganic ligands are also used for complexing and stabilizing the metal ion over a wide pH range. The ligands studied include, but are not limited to, citrate, oxalate, edetic acid (EDTA), humic acids, and ethylenediamine succinic acid (EDDS). The catalysts based on different metal elements and ligands were developed to improve the degradation of organics, promote the Fe^{3+}/Fe^{2+} redox cycle and decrease the sludge production [59, 60].

5. Reuse of the iron-containing sludge after Fenton reaction

A large amount of ferric sludge generated from the Fenton treatment. The practical applications of the Fenton process are limited, mainly because of neutralization after oxidation. The discharge of ferric sludge easily causes secondary pollution because of residual organics adsorbed and accumulated in ferric sludge from treated wastewater. The disadvantage is, therefore, the main obstacle limiting the development and application of the Fenton process [61].

Two approaches have been studied to minimize the production of sludge as a by-product of the Fenton process, including heterogeneous catalysts and the reuse of the iron-containing sludge [61]. However, the catalytic activity is usually weakened after repetitive use due to active iron leaching [62] or the decay of active catalytic sites [63]. Recently, the reuse of iron-containing Fenton sludge has been drawing increasing interest from researchers world-wide.

5.1 Fenton-like reactions using iron-containing sludge

The iron-containing Fenton sludge was used as an iron source for the synthesis of ferrite catalysts that have drawn much more attention due to their potential application in the fields of catalysis in the Fenton process.

Zhang et al. [64] proposed a novel method for the reuse of Fenton sludge in the synthesis of nickel ferrite particles ($NiFe_2O_4$). In phenol degradation with H_2O_2, $NiFe_2O_4$ alone, and $NiFe_2O_4$—H_2O_2, the phenol removal was as high as $95 \pm 3.4\%$. However, the phenol removal efficiencies were as low as $5.9 \pm 0.1\%$ and $13.5 \pm 0.4\%$ in H_2O_2 and $NiFe_2O_4$ alone, respectively. The leaching of iron ion from heterogeneous ferrite catalysts under the acid conditions is a common phenomenon. The leached iron amounted to $6.3 \pm 0.2\%$ of total iron, and the recovery ratio of $NiFe_2O_4$ catalyst in this study was found to be $97.1 \pm 1.7\%$ [64]. Notably, a rapid electron exchange between Ni^{2+} and Fe^{3+} ions in the $NiFe_2O_4$ structure could accelerate the conversion of Fe^{3+} to Fe^{2+}, which was beneficial for the Fenton reaction. In addition, the Fe^{3+} on the surface of $NiFe_2O_4$ particles and the leaching of iron ion from $NiFe_2O_4$ could also react with the H_2O_2 to induce Fenton reaction.

Therefore, phenol could be effectively removed [64]. Roonasi and Nezhad [65] compared the catalytic activity of nano ferrite M-Fenton sludge (M = Cu, Zn, Fe, and Mn), and $CuFe_2O_4$ achieved the best performance. Based on the previous studies, a new catalyst Cu_2O—$CuFeC_2O_4$ was synthesized by co-sedimentation. Compared with $CuFeC_2O_4$, Cu_2O—$CuFeC_2O_4$ improved the phenol removal. The phenol removal as high as $97.3 \pm 0.4\%$, and the leached iron amounted to 4.77% of total iron in Cu_2O—$CuFeC_2O_4$—H_2O_2 reaction. The superior catalytic performance was mainly due to the synergistic effect of both Cu^+ and Cu^{2+} as well as Fe^{2+}/Fe^{3+} redox pairs [66]. An electron bridge was formed between Cu^+ and Fe Fe^{3+}, which accelerates the formation of Fe^{2+} species in order to boost the reaction rate [66].

5.2 Regenerated the iron-containing sludge by the electrochemical process

Fenton process was employed to treat synthetic dye wastewater with a supply of Fe^{2+} electrolytically generated from iron-containing sludge [67]. The concentration of Fe^{2+} increases linearly (r^2 of 0.94) with increasing electrolysis time, but the amount of total iron provided is enough and not the limiting factor for the electrogeneration of Fe^{2+}. In electro-Fenton reaction, Fe^{3+} and O_2 were reduced to Fe^{2+} and H_2O_2 at the same time on the cathode. So, there exists competition between Fe^{2+} and H_2O_2 production. In order to eliminate the competition and decrease the chemical cost of H_2O_2, hypochlorous acid (HOCl) was instead of H_2O_2 [68]. Two iron sludge reuse modes were examined to treat 1,4-dioxane in this study: sequencing batch mode and separation batch mode. The current efficiency (CE) in the electrolytic cell is related to the initial iron concentration, the initial iron species, and operation pH. Fe^{3+} ions were perceived to be more suitable for use as the initial iron species in the electrochemical Fenton-type process, where the CE was found independent of the Fe^{3+} concentration. Compared with the sequencing batch mode, the iron recovery ratio was higher in the separation batch mode. Therefore, the separation batch mode is relatively suitable for iron sludge reuse for both the CE and the iron recovery rate [68].

5.3 Regenerated the iron-containing sludge by the thermal method

In order to remove the residual organics adsorbed in the waste sludge and minimize the sludge production. Baking the sludge was a common method [69, 70]. A higher the baking temperature led to less accumulation of organics in sludge.

The iron-containing sludge catalyst showed more activity (superior TPh, COD, and TOC removals) but higher leaching iron and less adsorption at high baking temperature. Also, the BOD_5/COD ratio was dropped by less than 50% [70]. However, the thermal methods will increase the overall cost and operational difficulty. Fe^{3+}/Fe^{2+} redox cycle cannot be effectively realized in thermal system, thus, decreasing the catalytic ability of the iron-containing sludge catalyst.

5.4 Regenerated the iron-containing sludge by reduction

Some organic ligands (such as EDTA, EDDS) are used for complexing and regenerating iron-containing sludge to promote the Fe^{3+}/Fe^{2+} redox cycle. These organic ligands effectively inhibited the precipitation of Fe^{3+}. However, organic ligands such as EDTA were difficult to b biodegraded, which remained in the water and caused the second pollution. These organic ligands cannot efficiently promote the Fe^{3+}/Fe^{2+} redox cycle, so reductants were required to help the Fe^{3+} reduction.

Organic reductants such as ascorbic acid, glutamic acid, and catechol were usually used in the Fenton-based process to accelerate the Fe^{3+}/Fe^{2+} cycle, enhance the performance of Fenton reaction and expand the range of operation pH. On the other hand, organic reductants can react with ·OH. Because the selectivity of the reaction between reductants and ·OH is different, the degradation efficiency of pollutants is different.

The investigation evaluated the efficacy of the ferric oxy-hydroxide sludge continuous reuse in the Fenton-based treatment of landfill leachate in a sequencing batch reactor with and without the addition of supplementary ferrous iron [71]. The mechanism of the ferric oxy-hydroxide sludge-activated hydrogen peroxide oxidation in the presence of strong complexing and reducing agents was proposed.

Three iron-dissolution mechanisms could be distinguished: protonation, complexation, and reduction. First, in the case of the H_2O_2/sludge system, the probability of iron dissolution by protonation is objectively high at favorable acidic conditions applied. Second, the dissolution of iron by complexation involves the attachment of a complexing ligand onto the ferric oxy-hydroxide surface; humic and fulvic acids, main subclasses in landfill leachate, contain a high density of functional groups, which can conjugate iron ions to form ion-ligand complexes. Third, these complexing ligands were partially dissolved at pH 3, and bound to the protonated OH-group, thus ending in ultimate decomposed of the Fe^{3+}-ligand complex into the bulk solution. Last, both humic and fulvic acids, as effective reductants, reduced Fe^{3+} via electron transfer mechanism [72].

Besides, the addition of a Fe^{2+} activator to the ferric oxy-hydroxide sludge-activated Fenton-based systems increases the total phenols removal rate and reuse cycles. The supplementation of the Fe^{2+} activator could ensure the full utilization of the oxidant.

Organics as quinone- and hydroquinone-structure compounds (such as tannic acid, lignin, phenol, and among others) may assist the Fenton oxidation by reducing Fe^{3+} to Fe^{2+} The loss iron can be used by these organics to form complexes and reduced in the H_2O_2/sludge cycle. Comparing the reducing efficiency of iron-containing sludge by the quinone-structure organics, it is found that H_2O_2/sludge/TN system gave the best performance after adding the quinone-structure organics. Tannins are considered to be strong metal-chelating and reducing agents [73]. They exhibit anti-oxidation (act as ·OH scavengers) and pro-oxidation (promote ·OH generation in the presence of transition metals) properties in biological systems (living organisms).

A comprehensive study of the catalytic performance of Fe^{3+} in the presence of tannic acid during the Fenton-based treatment of 2,4,6-trichlorophenol (TCP) was performed [74]. The addition of TN significantly promoted the degradation of TCP. H_2O_2 dose and ferric sludge load for water treatment in the presence of TN should be optimized to balance Fe^{3+} reduction to boost the Fenton reaction and ·OH scavenging to remove TN from water. The Fe^{3+} reductive mechanism by tannic acid incorporated tannic acid-Fe^{3+} complex formation and decay through an electron transfer reaction to form Fe^{2+}.

TN availability in wastewater allows for the reuse of non-regenerated ferric sludge for Fenton-based oxidation without any supplementary Fe^{2+}. The positive effect of TN in the Fe^{3+}/H_2O_2 system can be explained by its monomer of gallic acids. Gallic acid complexed with Fe^{3+} to form two complexes that protonated $[Fe(LH)]^{2+}$ and deprotonated $[Fe(L)]^{+}$ from pH 1.0 to 3.0. These complexes were decomposed into Fe^{2+} and the quinone group by electron transfer, while Fe^{2+} reacted with H_2O_2 to accelerate the TCP removal [74].

The reuse methods of iron-containing sludge have been successfully proved with synthetic aqueous solutions of highly toxic compounds, such as phenols and chlorophenols. However, the decontamination of real wastewater by these enhanced technologies has been so far scarcely performed. The main reason is that real wastewater contains many organic pollutants with high toxicity and low biodegradability. These pollutants cannot completely mineralize. Organic pollutants will be continuously accumulated in reused iron-containing sludge that produced from treating real wastewater by the Fenton process. The phenomenon causes the low efficiency of the Fenton reaction.

6. Conclusions

During the last few years, many research efforts have been made toward the improvement of the Fenton process. Hybrid methods such as photo-Fenton, electro-Fenton, and sono-Fenton are not economically viable techniques to degrade large volumes of effluent disposed of by the industries. Most experimental studies have been conducted at the laboratory scale; thus, a more detailed investigation is required for the Fenton process to be considered feasible for industrial treatment plants. Further research on the advancement of the Fenton process is needed to demonstrate the economic and commercial feasibility of this process.

Although heterogeneous catalysts demonstrate considerable advances for the elimination of contaminants, there are still drawbacks related to the low oxidation rates, which appeared when pH values above four along with iron leaching, leading to an increase in the H_2O_2 consumption. Future studies should address the stability of the process for broader operational conditions to avoid metal leaching of into the reaction solution and their negative effects on the environment. These combined methods and heterogeneous systems expected to reduce the production of Fenton sludge. However, the high cost of combined methods, the leaching of active iron, and the decay of active catalytic sites should limit the reduction of Fenton sludge.

At present, regenerating the Fenton sludge is crucial for future studies. The use of Fenton sludge as a catalyst for the Fenton process has been tested mainly after thermal regeneration and subsequent re-dissolution of iron-containing solids by acid, chemical regeneration with reducing agent, and electrochemical reduction. The decontamination of real effluents by these enhanced technologies has been so far scarcely performed. The main reason is that real wastewater contains many complex organic pollutants with high toxicity and low biodegradability. These pollutants cannot completely mineralize. Significant attention should be devoted in the future on the development of rate expressions (based on reaction mechanisms), identification of reaction intermediates, identification of scale-up parameters, and cost-effectiveness analysis.

Acknowledgements

The work is financially supported by the China special S&T project on treatment and control of water pollution (2017ZX07402002).

Author details

Min Xu[1,2], Changyong Wu[1*] and Yuexi Zhou[1*]

1 State Key Laboratory of Environmental Criteria and Risk Assessment, Chinese Research Academy of Environment Sciences, Beijing, China

2 College of Water Science, Beijing Normal University, Beijing, China

*Address all correspondence to: changyongwu@126.com and zhouyuexi@263.net

IntechOpen

© 2020 The Author(s). Licensee IntechOpen. Distributed under the terms of the Creative Commons Attribution - NonCommercial 4.0 License (https://creativecommons.org/licenses/by-nc/4.0/), which permits use, distribution and reproduction for non-commercial purposes, provided the original is properly cited.

References

[1] Babuponnusami A, Muthukumar K. A review on Fenton and improvents to the Fenton process for wastewater treatment. Journal of Environmental Chemical Engineering. 2014;**2**:557-572

[2] Lucas MS, Dias AA, Sampaio A, Amaral C, Peres JA. Degradation of a textile reactive Azo dye by a combined chemical-biological process: Fenton's reagent-yeast. Water Research. 2007;**41**: 1103-1109

[3] Huang CP, Dong C, Tang Z. Advanced chemical oxidation: Its present role and potential future in hazardous waste treatment. Waste Management. 1993;**13**(5-7):361-377

[4] Oturan AA, Aaron JJ. Advanced oxidation processes in water/ wastewater treatment: Principles and applications-a review. Critical Reviews in Environmental Science and Technology. 2014;**44**:2577-2641

[5] Gallard H, De Laat J. Kinetic modeling of Fe(III)/H_2O_2 oxidation reactions in dilute aqueous solution using atrazine as a model organic compounds. Water Research. 2000;**34**: 3107-3116

[6] Bigda RJ. Consider Fenton's chemistry for wastewater treatment. Chemical Engineering Progress. 1995;**91**: 62-66

[7] Duarte F, Maldonado-Hódar FJ, Madeira LM. Influence of the characteristics of carbon materials on their behavior as heterogeneous Fenton catalysts for the elimination of the azo dye Orange II from aqueous solutions. Applied Catalysis B: Environmental. 2011;**103**:109-115

[8] Yuan SH, Gou N, Alshawabkeh AN, Gu AZ. Efficient degradation of contaminants of emerging concerns by a new electro-Fenton process with Ti/MMO cathode. Chemosphere. 2013; **93**:2796-2804

[9] Wang NN, Zheng T, Zhang GS, Wang P. A review on Fenton-like processes for organic wastewater treatment. Journal of Environmental Chemical Engineering. 2016;**4**:762-787

[10] Matta R, Hanna K, Chiron S. Fenton-like oxidation of 2,4,6-trinitrotoluene using different iron minerals. The Science of the Total Environment. 2007;**385**:242-251

[11] Shinya A, Bergwall L. Pyrite oxidation: Review and prevention practices. Journal of Vertebrate Paleontology. 2007;**27**:145A-145A

[12] Babuponnusami A, Muthukumar K. Removal of phenol by heterogeneous photo electro Fenton-like process using nano-zero valent iron. Separation and Purification Technology. 2012;**98**: 130-135

[13] Dehghani M, Shahsavani E, Farzadkia M, Samaei MR. Optimizing photo- Fenton like process for the removal of diesel fuel from the aqueous phase. Journal of Environmental Health Science and Engineering. 2014;**12**:1-17

[14] Pouran SR, Aziz ARA, Daud WMAW. Review on the main advances in photo-Fenton oxidation system for recalcitrant wastewaters. Journal of Industrial and Engineering Chemistry. 2015;**21**:53-69

[15] Alfaya E, Iglesias O, Pazos M, Sanroman MA. Environmental application of an industrial waste as catalyst for the electro-Fenton-like treatment of organic pollutants. RSC Advances. 2015;**5**:14416-14424

[16] Brillas E, Sires I, Oturan MA. Electro-Fenton process and related electrochemical technologies based on

Fenton's reaction chemistry. Chemical Reviews. 2009;**109**:6570-6631

[17] Wang CK, Shih YH. Degradation and detoxification of diazinon by sono-Fenton and sono-Fenton-like processes. Separation and Purification Technology. 2015;**140**:6-12

[18] Carta R, Desogus F. The enhancing effect of low power microwaves on phenol oxidation by the Fenton process. Journal of Environmental Chemical Engineering. 2013;**1**:1292-1300

[19] Haber F, Weiss J. The catalytic decomposition of hydrogen peroxide by iron salts. Proceedings of the Royal Society of London, Series A. 1934; **147**(861):332-351

[20] Bray WC, Gorin MH. Ferryl ion, a compound of tetravalent iron. Journal of the American Chemical Society. 1932; **54**(5):2124-2125

[21] Neyens E, Baeyens J. A review of classic Fenton's peroxidation as an advanced oxidation technique. Journal of Hazardous Materials. 2003;**B98**(1-3): 33-50

[22] Pignatel JJ, Oliveros E, Mackay A. Advanced oxidation processes for organic contaminant destruction based on the Fenton reaction and related chemistry. Critical Reviews in Environmental Science and Technology. 2006;**36**:1-84

[23] Duesterberg CK, Mylon SE, Waite TD. pH effects on iron-catalyzed oxidation using Fenton's reagent. Environmental Science and Technology. 2008;**42**:8522-8527

[24] Szpyrkowicz L, Juzzolino C, Kaul SN. A comparative study on oxidation of disperse dye by electrochemical process, ozone, hypochlorite and Fenton reagent. Water Research. 2001;**35**: 2129-2136

[25] Kavitha V, Palanivelu K. Destruction of cresols by Fenton oxidation process. Water Research. 2005;**39**:3062-3072

[26] Xu XR, Li XY, Li XZ, Li HB. Degradation of melatonin by UV, UV/ H_2O_2, Fe^{2+}/ H_2O_2 and UV/Fe^{2+}/H_2O_2 processes. Separation and Purification Technology. 2009;**68**:261-266

[27] Yoon J, Lee Y, Kim S. Investigation of the reaction pathway of ·OH radicals produced by Fenton oxidation in the conditions of wastewater treatment. Water Science and Technology. 2000; **44**:15-21

[28] Zepp RG, Faust BC, Hoigné J. Hydroxyl radical formation in aqueous reaction (pH 3-8) of iron(II) with hydrogen peroxide the photo-Fenton reaction. Environmental Science and Technology. 1992;**26**(2): 313-319

[29] Zuo Y, Hoigne J. Formation of hydrogen peroxide and depletion of oxalic acid in atmospheric water by photolysis of iron(III) –oxalato complexes. Environmental Science and Technology. 1992;**26**:1014-1022

[30] Katsumata H, Kaneco S, Suzuki T, Otha K, Yobiko Y. Photo-Fenton degradation of alachlor in the presence of citrate solution. Journal of Photochemistry and Photobiology A. 2006;**180**:38-45

[31] Bandala ER, Brito L, Pelaez M. Degradation of domoic acid toxin by UV-promoted Fenton-like processes in seaeater. Desalination. 2009;**45**: 135-145

[32] Giri AS, Golder AK. Chloramphenicol degradation in Fenton and photo-Fenton: Formation of Fe^{2+}-chloramphenicol chelate and reaction pathways. Industrial and Engineering Chemistry Research. 2014;**53**: 16196-16203

[33] Xiao D, Guo Y, Lou X, Fang C, Wang Z, Liu J. Distinct effects of oxalate versus malonate on the iron redox chemistry: Implications for the photo-Fenton reaction. Chemosphere. 2014; **103**:354-358

[34] Mohajeri S, Abdul Aziz H, Hasnain Isa M, Ali Zahed M, Adlan MN. Statistical optimization of process parameters for landfill leachate treatment using electro-Fenton technique. Journal of Hazardous Materials. 2009;**176**:749-758

[35] Guinea E, Arias C, Cabot PL, Garrido JA, Rodríguez RM, Centella F, et al. Mineralization of salicylic acid in acidic aqueous medium by electrochemical advanced oxidation processes using platinum and boron-doped diamond as anode and cathodically generated hydrogen peroxide. Water Research. 2008;**42**: 499-511

[36] Panizza M, Cerisola G. Electro-Fenton degradation of synthetic dyes. Water Research. 2009; **43**:339-344

[37] Pipi ARF, De Andrade AR, Brillas E, Sirés I. Total removal of alachlor from water by electrochemical processes. Separation and Purification Technology. 2014;**132**:674-683

[38] Ranjit PJD, Palanivelu K, Lee CS. Degradation of 2,4-dichlorophenol in aqueous solution by sono-Fenon method. Korean Journal of Chemical Engineering. 2008;**25**:112-117

[39] Ioan I, Wilson S, Lundanes E, Neculai A. Comparison of Fenton and sono-Fenton bisphenol A degradation. Journal of Hazardous Materials. 2007; **142**:559-563

[40] Chakma S, Moholkar VS. Physical mechanism of sono-Fenton process. AICHE Journal. 2013;**59**:4303-4313

[41] Neppolian B, Jung H, Choi H, Lee JH, Kang JW. Sonolytic degradation of methyl tert-butyl ether: The role of coupled Fenton process and persulphate ion. Water Research. 2002;**S36**: 4699-4708

[42] Wu CY, Zhou YX, Sun QL, Fu LY, Xi HB, Yu Y, et al. Appling hydrolysis acidification-anoxic-oxic process in the treatment of petrochemical wastewater: Form bench scale reactor to full scale wastewater plant. Journal of Hazardous Materials. 2016;**309**:185-191

[43] Badawy MI, Wahaab RA, EI-Kalliny AS. Fenton-biological treatment processes for the removal of some pharmaceuticals from industrial wastewater. Journal of Hazardous Materials. 2009;**167**:567-574

[44] Martinez NSS, Fernandez JF, Segura XF, Ferrer AS. Pre-oxidation of an extremely polluted industrial wastewater by the Fenton's reagent. Journal of Hazardous Materials. 2003; **B101**:315-322

[45] Vergili I, Gencdal S. Applicability of combined Fenton oxidation and nanofiltration to pharmaceutical wastewater. Desalination and Water Treatment. 2015;**56**:3501-3509

[46] Abdel-Shafy HI, EI-Khateeb MA, Mansour MSM. Treatment of leather industrial wastewater via combined advanced oxidation and membrane filtration. Water Science and Technology. 2016;**23**:26-43

[47] Blanco J, Torrades F, De la Varga M, García-Montano J. Fenton and biological-Fenton coupled processes for textile wastewater treatment and reuse. Desalination. 2011;**286**:394-399

[48] Ioannou-Ttofa L, Michael-Kordatou I, Fattas SC, Eusebio A, Ribeiro B, Rusan M, et al. Treatment efficiency and economic feasibility of biological oxidation membrane filtration and

separation processes, and advanced oxidation for the purification and valorization of olive mill wastewater. Water Research. 2017;**114**: 1-13

[49] Mylon SE, Sun QA, Waite TD. Process optimization in use of zero valent iron nanoparticles for oxidative transformations. Chemosphere. 2010; **81**:127-131

[50] Chen Q, Wu P, Dang Z, Zhu N, Li P, Wu J, et al. Iron pillared vermiculite as a heterogeneous photo-Fenton catalyst for photocatalytic degradation of azo dye reactive brilliant orange X-GN. Separation and Purification Technology. 2010;**71**: 315-323

[51] Lan Q, Li F, Sun CX, Liu CS, Li XZ. Heterogeneous photodegradation of pentachlorophenol and iron cycling with goethite, hematite and oxalate under UVA illumination. Journal of Hazardous Materials. 2010; **174**:64-70

[52] Fu FL, Wang Q, Tang B. Effective degradation of Cl acid red 73 by advanced Fenton process. Journal of Hazardous Materials. 2010;**174**:17-22

[53] Pagano M, Volpe A, Lopez A, Mascolo G, Ciannarella R. Degradation of chlorobenzene by Fenton-like processes using zero-valent iron in the presence of Fe^{3+} and Cu^{2+}. Environmental Technology. 2011;**32**: 155-165

[54] Pecci L, Montefoschi G, Cavallini D. Some new details of the copper-hydrogen peroxide interaction. Biochemical and Biophysical Research Communications. 1997;**235**:264-267

[55] Choi K, Lee W. Enhanced degradation of trichloroethylene in nano-scale zero-valent iron Fenton system with Cu(II). Journal of Hazardous Materials. 2012;**211**:146-153

[56] Hermanek M, Zboril R, Medrik I, Pechousek J, Gregor C. Catalytic efficiency of iron (III) oxides in decomposition of hydrogen peroxide: Competition between the surface area and crystallinity of nanoparticles. Journal of the American Chemical Society. 2007;**129**:10929-10936

[57] Matta R, Hanna K, Kone T, Chiron S. Oxidation of 2,4,6-trinitrotoluene in the presence of different iron-bearing minerals at neutral pH. Chemical Engineering Journal. 2008;**144**:453-458

[58] Hanna K, Kone T, Medjahdi G. Synthesis of the mixed oxides of iron and quartz and their catalytic activities for the Fenton-like oxidation. Catalysis Communications. 2008;**9**: 955-959

[59] Song Z, Wang N, Zhu LH, Huang AZ, Zhao XR, Tang HQ. Efficient oxidative degradation of triclosan by using enhanced Fenton-like process. Chemical Engineering Journal. 2012; **198-199**:379-387

[60] Huang WY, Brigante M, Wu F, Mousty C, Hanna K, Mailhot G. Assessment of the Fe(III)-EDDS complex in Fenton-like processes: From the radical formation to the degradation of bisphenol A. Environmental Science and Technology. 2013;**47**:1952-1959

[61] Bolobajev J, Kattel E, Viisimaa M, Goi A, Trapido M, Tenno T, et al. Reuse of ferric sludge as an iron source for the Fenton-based process in wastewater treatment. Chemical Engineering Journal. 2014;**255**:8-13

[62] Ji F, Li C, Zhang J, Deng L. Efficient decolorization of dye pollutants with $LiFe(WO_4)_2$ as a reusable heterogeneous Fenton-like catalyst. Desalination. 2011;**269**(1-3):284-290

[63] Takbas M, Yatmaz HC, Bektas N. Heterogeneous photo-Fenton oxidation

of reactive azo dye solutions using iron exchanged zeolite as a catalyst. Microporous and Mesoporous Materials. 2008;**115**(3):594-602

[64] Zhang H, Liu JG, Ou CJ, Faheem SJY, Yu HX, Jiao ZH, et al. Reuse of Fenton sludge as an iron source for $NiFe_2O_4$ synthesis and its application in the Fenton-based process. Journal of Environmental Sciences. 2016;**5**:1-8

[65] Roonasi P, Nezhad AY. A comparative study of a series of ferrite nanoparticles as heterogeneous catalysts for phenol removal at neutral pH. Materials Chemistry and Physics. 2016; **172**:143-149

[66] Faheem M, Jiang XB, Wang LJ, Shen JY. Synthesis of Cu_2O-$CuFe_2O_4$ microparticles from Fenton sludge and its application in the Fenton process: The key role of Cu_2O in the catalytic degradation of phenol. RSC Advances. 2018;**8**:5740-5748

[67] Li CW, Chen YM, Chiou YC, Liu CK. Dye wastewater treated by Fenton process with ferrous ions electrolytically generated from iron-containing sludge. Journal of Hazardous Materials. 2007;**144**: 570-576

[68] Kishimoto N, Kitamura T, Kato M, Otsu H. Reusability of iron sludge as an iron source for the electrochemical Fenton-type process using Fe^{2+}/HOCl system. Water Research. 2013;**47**: 1919-1927

[69] Cao GM, Sheng M, Niu WF, Fei YL, Li D. Regeneration and reuse of iron catalyst for Fenton-like reactions. Journal of Hazardous Materials. 2009; **172**:1446-1449

[70] Rossi AF, Martins RC, Quinta-Ferreira RM. Reuse of homogeneous Fenton's sludge from detergent industry as Fenton's catalyst. Journal of Advanced Oxidation Technologies. 2013;**16**(2):298-305

[71] Kattel E, Trapido M, Dulova N. Treatment of landfill leachate by continuously reused ferric oxyhydroxide sludge-activated hydrogen peroxide. Chemical Engineering Journal. 2016;**304**:646-654

[72] Voelker BM, Sulzberger B. Effects of fulvic acid on Fe(II) oxidation by hudrogen peroxide. Environmental Science and Technology. 1996;**30**: 1106-1114

[73] Bolobajev J, Trapido M, Goi A. Role of organic wastewater constituents in iron redox cycling for ferric sludge reuse in the Fenton-based treatment. International Journal of Chemical, Molecular, Nuclear, Materials and Metallurgical Engineering. 2016;**10**(4): 352-357

[74] Bolobajev J, Trapido M, Goi A. Interaction of tannic acid with ferric iron to assist 2,4,6-trichlorophenol catalyst decomposition and reuse of ferric sludge as a source of iron catalyst in Fenton-based treatment. Applied Catalysis B: Environmental. 2016;**187**: 75-82

Chapter 5

Application of Ferrate for Advanced Water and Wastewater Treatment

Ansaf V. Karim, Sukanya Krishnan, Lakshmi Pisharody and Milan Malhotra

Abstract

Treatment of recalcitrant organics and inorganics present in wastewater is a major challenge. Conventional biological treatments alone are not capable of removing these toxic compounds from wastewater. To overcome these problems, advanced oxidation processes (AOPs) have been used to completely mineralize or transform the organics into simpler compounds, which can then be treated through biological processes. However, conventional AOPs result in the generation of byproducts, which are known to have higher toxicity. Among various AOPs, ferrate has been gaining popularity because of its advantages such as high oxidation potential, no byproduct formation and also non-toxic end products. The end product generated also acts as a coagulant, which thereby enhances the removal efficiency. In the present chapter, the chemical properties, preparation methods and the factors affecting the stability of ferrate were evaluated based on literature. Further, ferrate oxidation as a potential method for the treatment of both organic and inorganic pollutants in drinking and real wastewater is discussed.

Keywords: ferrate, oxidation process, disinfection, iron chemistry, synthesis

1. Introduction

Scientific discoveries and development in pharmaceuticals, pesticides, personal care products (PCP) and basic sanitation have improved the lifestyle and living standards of humans across the globe. However, these developments come with an environmental cost. In recent past, environmental and human health risks associated with traces of pharmaceutical, pesticides and microplastics (from PCP) have been studied by various researchers. Due to the transformation and interaction of these compounds with the environment, they can enhance the resistance of microbes, thus making them resistant to antibiotics. Furthermore, some of the pesticides have been proven to have a carcinogenic effect. The microplastics generated from PCP have now been reported to be the part of the food chain mostly from sea salts and fish. With these environmental challenges, there has been a sharp increase in the research for the treatment and removal of these emerging contaminants.

Various research groups have used different physicochemical techniques such as adsorption, Fenton treatment, zero-valent iron, ultraviolet (UV)-catalytic, electro-oxidation and wet oxidation [1–3] for the sustainable treatment of these contaminants. The techniques mentioned above have their advantages regarding their nonselective behaviour, high removal efficacy, and simpler byproducts. The higher cost associated with AOPs is a major limitation for field applicability.

From the past two decades, advanced oxidation processes (AOPs) have been gaining popularity for the treatment of water and wastewater contaminants. AOPs use hydrogen peroxide and ozone, among other chemicals that have high oxidation potential. These chemicals can oxidize organic molecules to simpler molecules and themselves get reduced. During AOPs, in situ hydroxyl radicals are formed, which

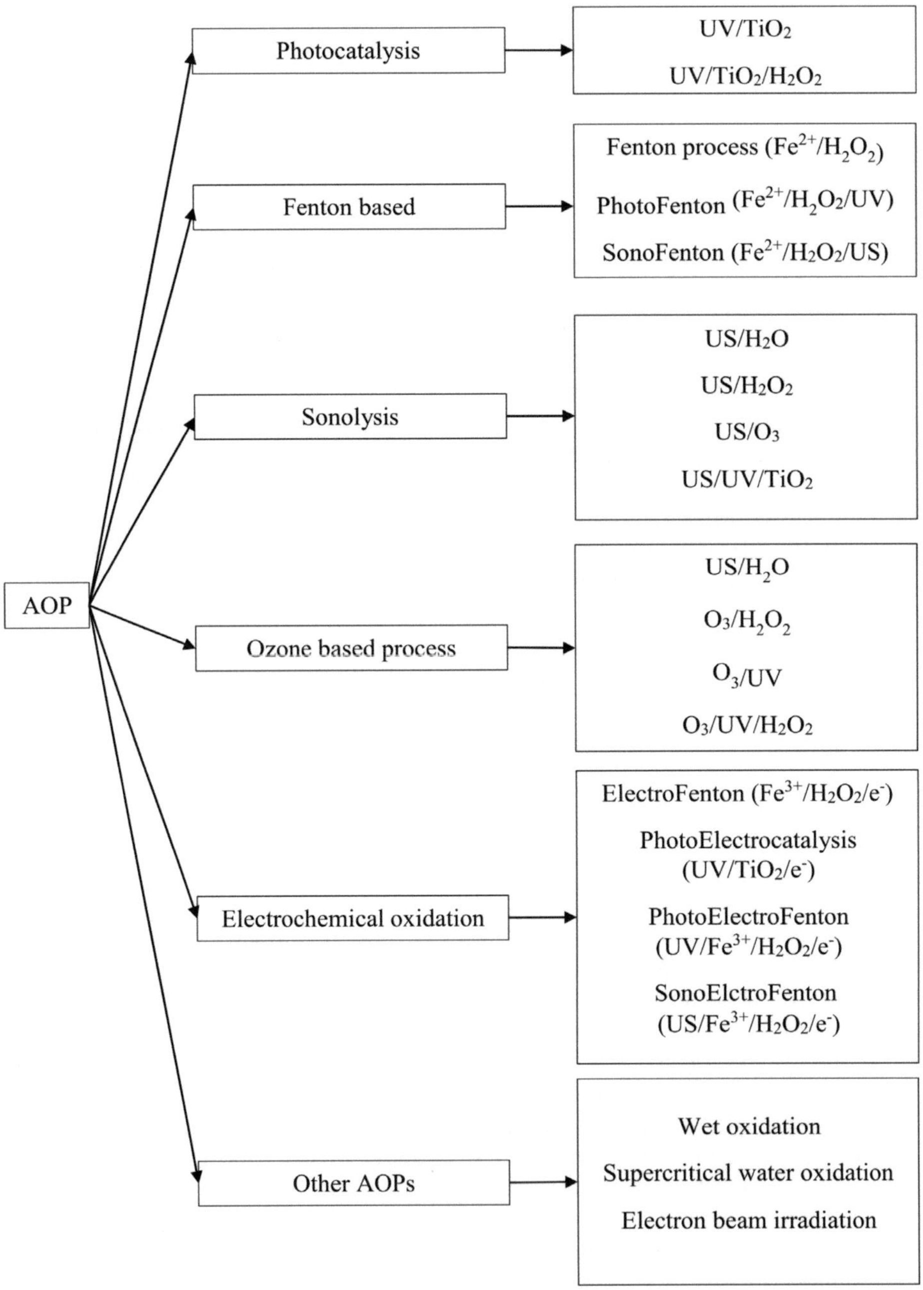

Figure 1.
Broad category of AOPs [4].

further reacts with organic molecules. The organic molecule may undergo various reactions such as partial cleavage, dechlorination, fragmentation or complete mineralization, depending on the chemical structure of molecule and reaction condition. The broad category of AOPs is shown in **Figure 1**.

Innovative technologies using alternative sources such as peroxides, ferrates, ozone, etc. can significantly improve the wastewater treatment due to their higher selectivity and efficiency. Among them, ferrate [Fe(VI)] is a promising choice for environmental remediation. It does not produce any harmful byproducts during treatment and provides efficient degradation of organic, inorganics and microorganisms over a wide range of pH. This chapter provides a detailed review of chemical properties, preparation methods and factors affecting the stability of ferrates. Further, the chapter also reviews the oxidation potential of ferrates for the environmental remediation and disinfection.

2. Chemical structure and stability of ferrate

Typically iron (Fe) exists in two oxidation states either as ferrous (Fe(II)) or ferric (Fe(III)). However, under strong oxidizing conditions, a higher oxidation state (i.e. +IV, +V and +VI) of iron can be achieved, which is referred to as a ferrate. Based on the pE-pH diagram, which is also known as Pourbaix diagram of iron (**Figure 2**), we can visualize the presence of various forms of iron at different pH and under oxidative/reductive environment. A reducing environment is represented by low pE values; high pE values represent an oxidizing environment. Ferrates are predominating in the top region of the pE-pH diagram, whereas iron exists as Fe (0) at the bottom.

Among the various higher oxidation state, the +VI state of iron is comparatively stable and easier to produce. The literatures, based on XRD of solid ferrate, proposed a tetrahedral structure of ferrate (**Figure 3**). In ferrate, centrally placed Fe^{+6} is covalently bonded to four equivalent oxygen atoms at the corner of the tetrahedron.

In the aqueous phase, there are many factors (such as pH, temperature and ferrate concentration) on which the stability of ferrate depends. Ferrate ions and water molecules reacts to form ferric oxide (Fe_2O_3), oxygen gas, and hydroxyl ions (Eq. (1)). Due to the release of hydroxyl ions, the resulting pH of the solution is highly alkaline. The ferric oxide generated as an end product acts as a coagulant:

$$4FeO_4^{2-} + 4H_2O \rightarrow 2Fe_2O_3 + 3O_2 + 8OH^- \quad (1)$$

Under the acidic environment, ferrate is quite unstable and rapidly undergoes exothermic degradation. In aqueous solution of pH ~5, the ferrate (1 mM) was reported to be completely degraded within 7 min [5]. The stability of ferrate tends to improve with an increase in pH, as can be seen in **Figure 4**. For pH ~10, the ferrate concentration remained almost constant. At pH of above 10, the stability of ferrate also tends to deteriorate. For a ferrate concentration of 0.25 mM and a pH of 12, reductions of ~ 60% were reported within 10 min [6].

A diluted ferrate solution is reported to be more stable than a concentrated solution. The critical ferrate concentration is 0.025 M; a ferrate concentration higher than this (0.025 M) in solution tends to degrade rapidly. When comparing the stability of ferrate at different concentrations, it was reported that ~89% of ferrate concentration remained as ferrate for 0.02 and 0.025 M strength for 1 h. However, complete degradation was reported for solution with a 0.03 M concentration in that specific duration [7].

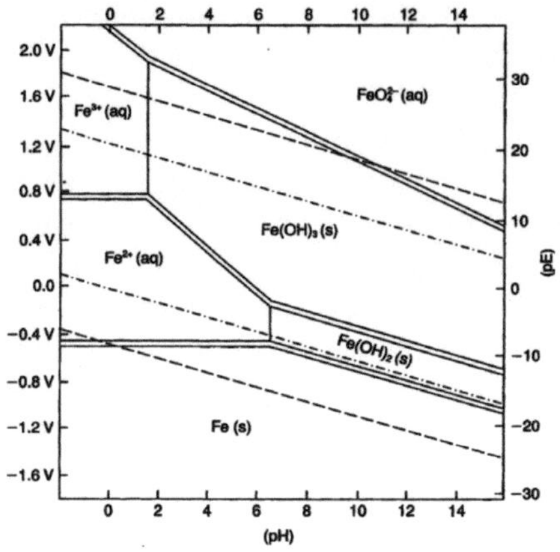

Figure 2.
Pourbaix diagram for iron (1 M) solution.

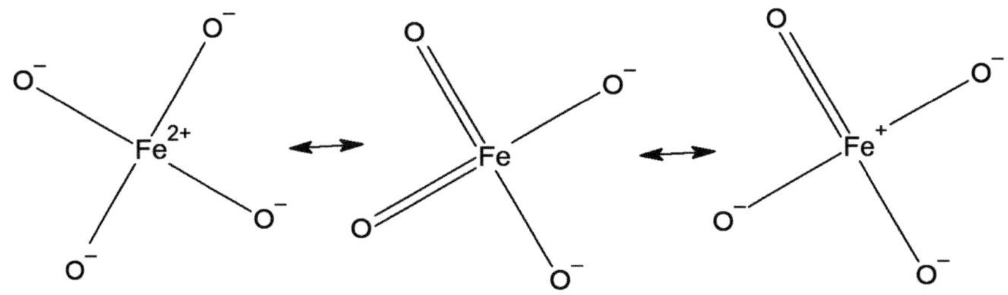

Figure 3.
Resonance hybrid structure of ferrate.

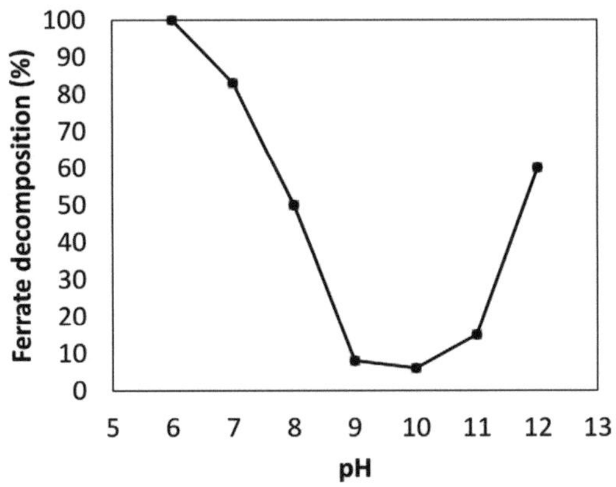

Figure 4.
Stability of ferrate in the aqueous phase at different pH [6].

Co-ions such as Ni(II) and Co(II) in 100 μM concentration were reported to increase the decomposition rate of ferrate (2 mM). However, at lower concentration, the influence of co-ion (0.1 μM) on ferrate decomposition was reported to diminish. No significant influence of other ions (Ba(II), Ca(II), Fe(III), Mg(II), Zn(II), Pb(II), Cu(II), Sr.(II)) and salts (KIO_4, $K_2O_9B_4$, K_3PO_4, Na_2WO_4, Na_2MoO_4, Na_2SiF_6) on ferrate degradation was reported [8].

The temperature of the aqueous phase can also influence the stability of ferrate. Ferrate solution (0.01 M) kept at 0.5°C was reported to remain stable

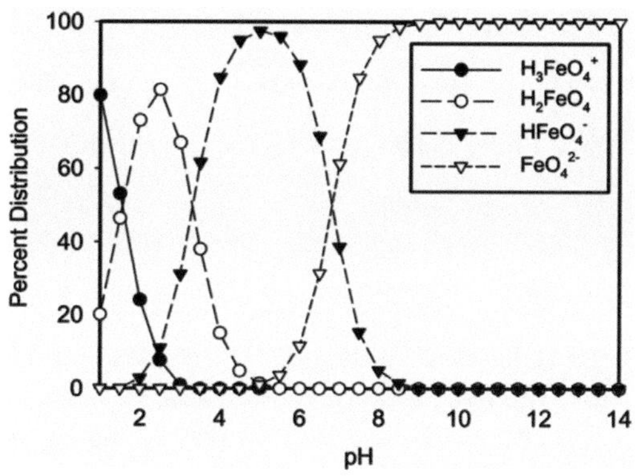

Figure 5.
Different species of ferrate in the aqueous phase [10].

(~2% reduction) even after 2 h. On the other hand, the ferrate concentration was found to be reduced by ~10% at a temperature of 25°C [9]. External factors such as light are not known to have any significant influence on the stability of ferrate (0.01 M) [9].

Depending on the pH of the solution, four different species of ferrate can exist, as shown in **Figure 5**. Initially, two unstable species of Fe(VI) (i.e. H_2FeO_4 and $HFeO_4^-$) were reported in phosphate buffer (0.2 M solution). Three protonated species of Fe(VI) were reported in phosphate/acetate buffer (0.025). Based on pk_a values, it can be seen that in the alkaline environment, most of the ferrate exist as FeO_4^{2-}. In neutral conditions, $HFeO_4^-$ is the dominating species:

$$H3FeO_4^+ \leftrightarrow H^+ + H_2FeO_4 \left(pK_a = 1.6\right) \quad (2)$$

$$H_2FeO_4 \leftrightarrow H^+ + HFeO_4^- \left(pK_a = 3.5\right) \quad (3)$$

$$HFeO_4^- \leftrightarrow H^+ + FeO_4^{2-} \left(pK_a = 7.3\right) \quad (4)$$

3. Synthesis of ferrate

3.1 Wet synthesis

The wet oxidation synthesis of ferrate involves oxidation of ferric to ferrate under high pH conditions. The solution containing ferrate obtained by this procedure is highly unstable, which thereby demands subsequent procedures of precipitation, washing and drying to obtain a stable solid product. Numerous methods have been tried to prepare a stable solid product containing ferrate(VI). However, the subsequent recovery of ferrate is difficult because of the increased solubility of Na_2FeO_4 in solution saturated with NaOH. The procedure involving the flow of chlorine gas through the ferric salt resulted in a solid product with 41.38% of Na_2FeO_4 [11, 12].

The method of production of ferrate by wet oxidation is known since the 1950s. The method involves a reaction between ferric chloride and sodium hypochlorite in the presence of an alkali such as sodium hydroxide. Further recovery of potassium ferrate is obtained by precipitation with potassium hydroxide. Even though a percentage recovery of potassium ferrate as high as 96% could be obtained, the maximum yield percentage obtained by continuous efforts was 75%. Earlier the recovery was poor as 10–15%. The basic reactions involved in the process are shown below [13, 14]:

$$Fe^{3+} + 3OH^- \rightarrow Fe(OH)_3 \tag{5}$$

$$2Fe(OH)_3 + 3NaClO + 4NaOH \rightarrow 2Na_2FeO_4 + 3NaCl + 5H_2O \tag{6}$$

$$2Na_2FeO_4 + KOH \rightarrow K_2FeO_4 + 2NaOH \tag{7}$$

3.2 Dry synthesis

The dry oxidation method for ferrate synthesis is a very old method. However, the method involves high risks as it can lead to an explosion at elevated temperature. Recent developments include synthesis of ferrate salt by calcination of a mixture of potassium peroxide and ferric oxide at a temperature ranging from 350 to 370°C. Another method involves oxidation of iron oxide with sodium peroxide at a temperature of 370°C, with a continuous flow of dry oxygen in a ratio of 4:1, yielding sodium ferrate. The product by this reaction results in a red-violet colored solution containing tetrahedral ion $FeO4^{2-}$ [11, 15, 16]. The % recovery of ferrate was not higher than 55% in this method [17]. The equation for the reaction is given below:

$$Fe_2O_3 + 3Na_2O_2 \rightarrow 2Na_2FeO_4 + Na_2O \tag{8}$$

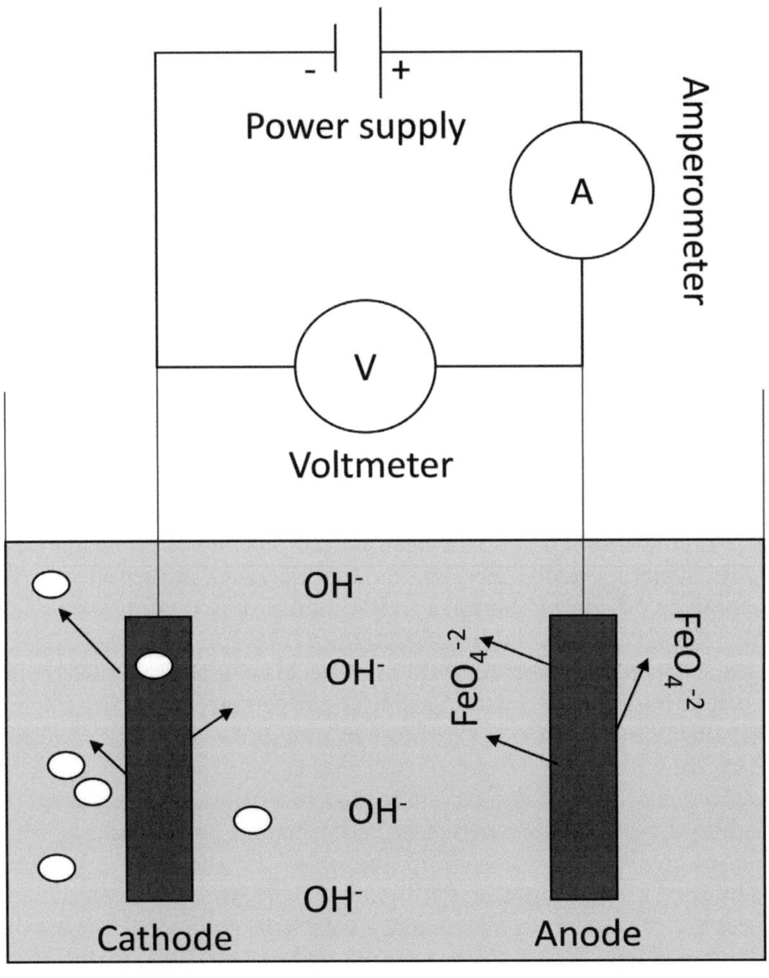

Figure 6.
Electrochemical cell for ferrate production [18].

3.3 Electrochemical synthesis

During the electrochemical synthesis of ferrate, anodic dissolution of iron takes place in a strongly alkaline solution. The current applied during the synthesis process oxidizes the iron to ferrate in the alkaline solution (KOH or NaOH) (**Figure 6**). The reactions at the anode and cathode are as follow:

$$\text{Reaction at anode}: Fe + 8OH^- \rightarrow FeO_4^{2-} + 4H_2O + 6e^- \quad (9)$$

$$\text{Reaction at cathode}: 3H_2O + 6e^- \rightarrow 3H_2\,(\uparrow) + 6OH^- \quad (10)$$

$$\text{Overall reaction}: Fe + 2OH^- \rightarrow FeO_4^{2-} + 3H_2 + 3H_2O \quad (11)$$

Factors such as anode composition, current density and strength of electrolyte govern the production of ferrate. The major advantage of electrochemical synthesis is its simplicity and no costly chemical requirement.

4. Application of ferrate in environmental remediation

The increasing occurrence of emerging pollutants such as pharmaceuticals, dyes, heavy metals and endocrine disrupting compounds, among others, in wastewater has generated a growing concern. Due to the recalcitrant nature of these compounds, they remain in the environmental matrices for longer duration and are not degraded naturally or by biological reactions. Ferrate as an oxidizing agent is a promising choice for the remediation of water and wastewater containing recalcitrant pollutants. They are able to oxidize compounds in a shorter time period when compared to permanganate and chromate [19]. The degradation potential of ferrate for water and wastewater treatment mainly depends upon the pH, initial ferrate concentration, presence of coexisting ions and temperature [20]. They can be easily reduced to insoluble and non-toxic Fe(III) species and can adsorb organic compounds, resulting in its removal [21]. In addition to the higher oxidizing ability of Fe(VI), other intermediate oxidation states of iron Fe(V) and Fe(IV) are claimed to help oxidation of organic and inorganic compounds [19]. The following section describes the application of ferrate as an oxidant for remediation of various pollutants.

4.1 Phenolic compounds

The removal of phenolic compounds from environmental matrices has attained considerable attention in recent years due to its recalcitrant and hazardous properties. The oxidative removal of these types of compounds from water and wastewater using ferrates Fe(VI) has been extensively studied by the researchers these days. Chen et al. [22] have examined the removal of four different types of phenolic compounds, such as 2-benzylphenol, phenol, chlorophene and 4-chlorophenol by Fe(VI) at pH 8. It was observed that the presence of chlorine and benzyl groups helps in the increased reactivity of Fe(VI) with phenolic compounds, and the degradation was observed to be in the order of chlorophene > 4-chlorophenol > 2-benzylphenol > phenol. They established the complete degradation pathway and various intermediates formed by mainly four pathways, such as hydroxylation of the benzene ring, C—C bond cleavage, chlorine atom substitution by a hydroxyl group and single-electron coupling.

A similar study of Chen et al. [23] demonstrated almost complete degradation of polychlorinated diphenyl sulphides (PCDPSs) within a short period of time

(~240 s) at pH 8. Also, they established the pathway of degradation and toxicity potential of products, which suggests that PCDPSs' oxidations occur at the S(II) moiety by ferrates and form oxidation products that are non-toxic to aquatic organisms. pH is one of the critical factors for oxidation reactions which also affects the stability and oxidation potential of ferrate ion [6]. However, high alkaline pH of liquid ferrates imparts an adverse effect on disinfection treatment by the formation of harmful disinfection byproducts [24]. Hence, detailed research should be carried out for understanding the effect of pH in the treatment of various recalcitrant pollutants with different ferrate concentrations.

The study conducted by Zhang et al. [25] demonstrated the use of Fe(VI) for the removal of one of the most renown endocrine disruptors, bisphenol-A, from water. They observed the complete degradation of BPA at Fe(VI)/BPA molar ratio of 3.0 and 1.5 mg L^{-1} BPA initial concentration within 30 minutes of oxidation under weak acidic pH. In order to enhance the degradation of various recalcitrant compounds, various homogeneous reducing agents such as sodium thiosulfate ($Na_2S_2O_3$) and sodium sulphite (Na_2SO_3) have been used in recent years. In the recent study by Sun et al. [26], the application of heterogeneous carbon nanotube along with ferrates for enhanced degradation of bromophenols. The reaction kinetics was found to followed second-order over a wide range of pH (6-10). It was found that the coupling of one-electron transfer mechanism caused the formation of some undesirable products, such as di-brominated phenoxyphenols and dihydroxylated biphenyls.

4.2 Dyes and dye wastewater

The application of Fe(VI) for removal of textile dyes is one of the promising technologies because it can act as an oxidant as well as a coagulant at the same time and it enhances the biodegradability property of dye. Li et al. [27] investigated the azo dye orange II removal by different types of Fe(VI), namely, potassium permanganate, potassium ferrate(VI) and the ferrate(VI) hypochlorite liquid mixture. It was observed that within 30 minutes, maximum discoloration of 95.6% was achieved with ferrate(VI) hypochlorite liquid mixture over a pH range of 3 to 11. Whereas, only 17.7% and 62 % of discoloration was observed respectively when only potassium permanganate and potassium ferrate(VI) were employed. However, one of the drawbacks of the above process is that residual hypochlorite in ferrate (VI) hypochlorite liquid mixture produces dangerous disinfection byproducts.

In another study, Xu et al. [28] utilized a highly stable composite ferrate(VI) solution produced by the chemical method with KOH at 65°C for the degradation of azo dye reactive brilliant red X-3B. The optimal conditions were found with dye concentration 0.08 mmol L^{-1}, pH = 8.4 and ferrate dosage 2.5 mg L^{-1}, giving discolouration of 99% after 20 min also; about 42% COD removal and 9% TOC removal were observed after 60 min. Also, the degradation pathway was theoretically established by the cleavage of by the C—N and N=N bond yielding muconic acid as the final reduced compound. In addition, the study by Turkay et al. [29] investigated the kinetics and mechanism of degradation of methylene blue by ferrate(VI). It was observed that about 96.82% methylene blue removal is achieved with initial methylene blue concentration of 50 mg L^{-1}, pH = 13.6 and ferrate dosage = 59.5 mg L^{-1} over 35 minutes of reaction time and followed second-order kinetics.

4.3 Pesticides

Fe(VI)-based oxidation processes are effective in improving the biodegradability of wastewater containing persistent organic pollutants such as pesticides.

In a study conducted by Zhu et al. [21], the ferrate-based pretreatment of wastewater containing a higher concentration of alachlor improved its biodegradability and resulted in the total removal of alachlor within 10 min under optimized conditions. A comparison of the oxidation property of peroxymonosulfate(PMS) and Fe (VI) and for the degradation of atrazine was studied by Wu et al. [30]. They have observed a significant enhancement in the degradation of atrazine in Fe(VI)/PMS system when compared to Fe(VI) or PMS alone.

In another study conducted by Chen et al. [31], the combined oxidation ability of the UV/Fe(VI) system for the removal of an organophosphorus pesticide profenofos was studied. They have observed that degradation of the compound was dependent on the pH and Fe(VI) dosage, and the degradation progressed through de-ethylation and de-propylation with subsequent C—O cleavage with the release of orthophosphate. The effect of operational parameters such as pH, Fe(VI) dosage and other water constituents (anions and cations) was studied while evaluating the potential of Fe(VI) oxidant on parathion (PTH) degradation [32]. Under the operation condition of pH 7 and Fe(VI): PTH 15:1, 99% of removal was observed, the presence of Fe^{3+}, HCO_3^-, Cu^{2+}, HA, Ca^{2+} decreased the removal rate, while no significant effect was observed for Cl^- and NO_3^-. In a similar study conducted by Liu et al. [33], they have concluded that the degradation of chlorpyrifos was mainly through C=O cleavage and hydroxyl substitution reactions.

4.4 Pharmaceuticals

The presence of pharmaceutically active compounds in the aquatic environment can cause serious health issues to the living organism. The strong oxidation potential of Fe(VI) and its ability to remove nitrogen- and sulfur-containing compounds had gained attention for the removal of pharmaceuticals from wastewater. The kinetics of removal of an antibacterial drug sulfamethoxazole (SMX) using potassium ferrate oxidant was studied by Sharma et al. [34]. They found that the rate constants are pH-dependent, and at higher concentrations of potassium ferrate (10 μm), the half-life of the compound was found to be 2 min at pH 7. In a similar kind of study, when the kinetics of removal of antiphlogistic drug ibuprofen (IBP) was studied, the degradation rates decreased with an increase in pH, which was related to protonation of Fe(VI) [35].

Another study conducted by Ma et al. [36] used sodium-potassium ferrate for the removal of tetracycline hydrochloride (TC) antibiotics at different pH and molar concentrations. Although maximum degradation was observed within a pH range of 9–10 and at a molar ratio between 1:1 and 1:10 [Fe(VI):TC], only 15% mineralization of the compound was observed due to the formation of stable intermediate products. The toxicity and kinetics of diclofenac removal by Fe(VI) under simulated disinfection conditions were studied by Wang et al. [37], and they have observed a rapid increase of the inhibition rate of the luminescent bacteria due to the formation of toxic intermediates. Also, the toxicity of Fe(VI)-treated wastewater containing pharmaceuticals was assessed on zebrafish embryo by Jiang et al. [38] and found that the level of toxicity decreased after the treatment.

The combination of Fe(VI) treatment with radiation for the removal of carbamazepine was studied by Wang et al. [39] for increasing the degradation efficiency and mineralization. They observed an increase in TOC removal during simultaneous treatment. Although the overall degradation rate of carbamazepine decreased, enhanced mineralization was observed with the sequential treatment of radiation, followed by Fe(VI), which may be due to a decrease in pH of the solution from 7.0 to 5.4, where the Fe(VI) oxidation was higher. The performance of electrogenerated ferrate(VI) and solid form of Fe(VI) were compared for the degradation of amoxicillin and ciprofloxacin [40]. By modeling the ferrate decomposition to understand its oxidation potential, they have concluded that the increased removal efficiency of compounds was due to the presence of a deprotonated form of ferrate ion.

In a study conducted by Yang et al. [41] for the removal of a mixture of emerging contaminants spiked in secondary effluent, only selective oxidation of electron-rich compounds was observed with Fe(VI)-based treatment process. Nevertheless, higher Fe(VI) exposure involved faster destruction of these recalcitrant compounds, proving the effectiveness of these treatments as an efficient tertiary treatment for broad-spectrum micropollutants in wastewater. In another study where mixture of four pharmaceuticals were treated with Fe(VI) (3 mg/L) all the compounds except bezafibrate were degraded effectively [42]. From this observation they have suggested that Fe(VI) is a selective oxidant, and the degradation efficiency will decrease if a carboxylic group is present in the compound.

4.5 Inorganic removal by ferrates

Fe(VI) became environmentally friendly oxidants due to its high oxidizing ability and did not produce any toxic compounds on decomposition. Therefore, the application of ferrates has been widely investigated among the researches for the removal of inorganics from water. Inorganic pollutants such as heavy metals, inorganic salts, trace elements and complexes of metals, among others, can also be removed by Fe(VI). There are basically two types of mechanisms in the ferrate oxidation of inorganic compounds, namely, one-electron transfer and two-electron transfer. Also, the degradation of inorganics is observed to be finished in milliseconds to seconds [19]. For instance, the study by Johnson and Hornstein [43] reported the complete oxidation of hydroxylamine to nitrous oxide by ferrates following a two-electron transfer oxidation mechanism, where the transfer occurs from either ferrates or water.

The application of ferrates for phosphate removal in municipal wastewater treatment plants is a feasible option and needed to be explored in detail. Lee and Zu [44] reported around 80% phosphate removal with a ferrate dose of 7.5 mg Fe L^{-1}. They investigated the performance of various forms of iron such as Fe(VI), Fe(II) and Fe(III) and observed that the efficiency of Fe(VI) is 1.5 times higher than Fe (III) because of the larger aggregate formation with Fe(VI) having higher specific surface. A similar study by Lee and Um [45, 46] utilized the combined use of Fe(III) and Fe(VI) for removal of arsenic by coagulation from river water samples and observed that 90% removal was achieved with an optimum dose of 2 mg L^{-1} Fe(VI) at pH 7.8. Also, potassium ferrate was used to remediate a solution containing a mixture of arsenic and antimony, where at first the compound was oxidized by Fe(VI) and then separated from the solution by the adsorption of its decomposition product [47]. In addition, the kinetics and mechanism of As(III) oxidation by Fe(VI) were studied by Lee et al. [45, 46], and they concluded that oxygen transfer mechanism in aqueous solution resulted in oxidation of As(III) to less toxic As(V).

Ferrate-based oxidation process was used for the removal of insoluble Cr(III) compounds from high-level radioactive tank waste, which was alkaline media [48]. $Fe_3O_4^{2-}$ concentration played a significant role in the Cr(III) removal; however, increasing the Fe(VI)/Cr(III) ratio greater than 10 reduced the removal efficiency. Ferrates are also capable of oxidizing organic metal complex species with its simultaneous removal from aqueous solutions [49, 50]. They are successfully used in industrial wastewater containing metal complexes, and treatment process efficiency depends on pH and Fe(IV) dosage. Among the different metal complexes, zinc cyanide (CN) found in the effluent wastewater of metal finishing industries had a significant environmental impact.

The effect of Zn(II) on the removal of cyanide by Fe(VI) was studied by Yngard et al. [51] at different alkaline pH conditions. Although they have observed a decrease in the oxidation rate of cyanide in the presence of Zn(II), complete removal was achieved by Fe(VI). Another ferrate-based oxidation study was conducted by Lee and Tiwari [52] for the simultaneous removal of CN, CN-Cu and CN-Ni. They have observed that in the CN-Cu and CN-Ni systems, rapid oxidation of CN was observed, while a relatively slower degradation of CN was observed in the mixed CN-Cu-Ni. In another study, Ni(II)-CN and Ni(II)-CN-ethylenediamine-tetraacetate (EDTA) solutions were removed using Fe(VI) as a function of pH ranging from 8 to 11 [53]. The presence of an organic ligand EDTA enhanced the removal of cyanide in Ni(II)-CN-EDTA, while only 60% removal was observed with Ni(II)-CN metal-cyanide complex. The stronger complexes formed due to the addition of EDTA into the Ni(II)-CN metal-cyanide complex reduced the amount of additional Fe(VI) required to oxidize cyanide in it. A similar study conducted by Pachuau et al. [54] assessed the suitability of ferrate(VI) for the simultaneous oxidation of Cu(II)-IDA and Zn(II)-IDA (iminodiacetic acid). It was observed that maximum removal Cu(II) or Zn (II) occurred at higher pH values.

4.6 Organic matter and wastewater

Fe(VI) can be effectively and efficiently used for removal of the organic matter present in wastewater. The stronger oxidizing property along with its disinfection and coagulation property makes Fe(VI) an ideal chemical for sewage treatment [55]. One of the main advantages of ferrate treatment is that it produces very less volume of sludge and thus makes sludge treatment much easier [56].

The study by Jiang et al. [12] investigated the applicability of ferrates for disinfection/coagulant and compared the performance with conventional disinfectants and coagulants such as sodium hypochlorite (NaOCl) and NaOCl combined with ferric sulphate or alum for the treatment of water and observed around 20% more removal with ferrates at a pH between 6 and 8. Among the three chemicals, ferric sulphate, potassium ferrate(VI) and aluminum sulphate used by Jiang et al. [55] for sewage treatment, maximum color removal and COD reduction were observed for potassium ferrate(VI). The dual properties of Fe(VI) species capable of oxidizing micropollutants and inorganics in wastewater were assessed for municipal wastewater treatment [57]. They have observed that Fe(VI) was efficient in oxidizing micropollutants and subsequently removing phosphate from a secondary treated effluent in a single treatment step.

In another study, natural organic matter (NOM) present in river and stream water was removed using potassium ferrate [58]. They have observed more than 70% removal of humic and fulvic acids by varying the Fe(VI) dosage, pH and temperature of the solution. The efficiency of ferrate and chlorine on removing organic matters present in secondary treated effluent was compared for two different wastewater treatment plants by Gombos et al. [59]. When comparing the bacterial inactivation, both Fe(VI) and chlorine were very effective even at their lower concentration, while a higher COD and TOC removal were observed for Fe (VI) in comparison to chlorine. The results showed that the higher oxidation capacity along with its coagulation and flocculation property of the reduced Fe(III) ions enhanced the wastewater treatment.

5. Ferrate as a disinfectant

Coagulation and disinfection are the two major unit operations that are important to ensure proper water quality. Disinfection has been known to be a tertiary treatment for the removal of pathogens. While coagulation destabilized the colloids and aggregates them, after which they can be settled and thereafter removed. Over the years, there has been rapid deterioration of the water quality due to increased water pollution. Hence, efficient chemical reagents and treatment techniques are required to meet stringent water quality standards. Various disinfection processes are used these days such as chlorination, UV irradiation and advanced oxidation processes [60].

The most common disinfection method used is chlorination. However, chlorine in the presence of organic substances such as humic acid is known to produce trihalomethanes (THM), which are known carcinogens. Ozonation results in the formation of N-nitrosodimethylamine (NDMA). While, presence of bromide ion also initiates formation of a potential carcinogen bromate [12, 13]. Thus, removing these disinfection byproducts is another challenge. An ideal chemical reagent would thus be a treatment reagent capable of carrying out both disinfections and should also be capable of oxidizing organic as well as inorganic pollutants such as heavy metals. Ferrate, with a redox potential of 2.2 V, higher than ozone (2.0 V), satisfies both requirements [61]. Ferrate is also considered as a "green" chemical. This is primarily because ferrate oxidation does not lead to the production of any harmful byproducts. The product of Fe(VI) oxidation is either ferric hydroxide or Fe(III) [14, 15].

5.1 Effect of ferrate ions on bacteria

Murmann and Robinson were the first to demonstrate the disinfection property of ferrate in the year 1974. The disinfection studies were conducted on two pure strains of non-recombining *Pseudomonas*, combining *Pseudomonas* and Missouri River Water. It was observed that complete killing of bacteria occurred at ferrate concentrations of 4, 50 and 100 ppm, respectively, for all the three tested samples. It was found that ferrate(VI) could kill 99.9% of microbes, such as *E. coli*, within 7 min of contact time and at a dose of 6 mg L^{-1} at a pH of 8. Also, the disinfection ability enhanced when the pH was below 8. For the same concentration of bacteria, a contact time of 17 min was required in order to achieve the same level of disinfection when the ferrate concentration was reduced to 2.4 mg L^{-1} [62].

The performance of ferrate as a disinfectant for drinking water was also compared with chlorine. Ferrate was found to achieve >$6\log_{10}$ inactivation of *E. coli* at a very low dose of 6 mg L^{-1}, while 10 mg L^{-1} of Cl_2 was required to achieve the same level of disinfection. Also, ferrate(VI) was found to work efficiently over a wide range of pH [12]. Similarly, for sewage treatment, potassium ferrate was found to achieve >4 $\log_{10}$ inactivation of bacteria, while alum sulphate and ferric sulphate were able to inactivate only 1 $\log_{10}$ and 1.5 $\log_{10}$ bacteria. Also, the dosage of ferrate required was lesser than that of alum sulphate and ferric sulphate [55]. Effectiveness of ferrate as a disinfectant was also studied on aerobic spore-bearing bacteria, which suggested that sodium ferrate is a powerful disinfectant of this group of bacteria. The study showed that sodium ferrate was a better indicator than even chlorine [63].

Ferrate was also found to inhibit the dehydrogenase activity by inactivating the —SH radical and thus inhibiting both the exogenous and endogenous respiration of bacteria *Sphaerotilus*. The study suggests that intermediates generated during the ferrate decomposition penetrate the cell wall of the bacterium and inactivate the

—SH radical which is necessary for dehydrogenase activity and thus respiration of the bacterium. *Sphaerotilus* is known to cause bulking in activated sludge [64]. Clostridia resistant to chlorination were also found to be killed by ferrate. *Bacillus cereus*, *Streptococcus bovis*, *Staphylococcus aureus*, *Shigella flexneri*, *Streptococcus faecalis* and *Salmonella typhimurium* were also found to be susceptible to ferrate disinfection [4, 9, 10]. However, the gram-positive organisms are found to be more resistant to the ferrate action compared to gram-negative organisms such as *E. coli*.

5.2 Effect of ferrate ions on viruses

Effectiveness of ferrate as a disinfectant on virus removal was checked on human surrogate viruses such as MS2, f2, Qβ and P22 bacteriophages [11–13]. Studies suggested that Fe(VI) was more efficient against naked icosahedral coliphages such as MS2 than gram-negative bacteria, *E. coli*. The k_i value for *E. coli* was reported to be 6.25×10^{-1} L/(mgFe × min) (pH 7.2, 25°C), while the k_i value for MS2 inactivation was 2.27 (±0.05) L/(mgFe × min). The faster kinetics for virus inactivation might be because of the denaturation of the genome or transformation of the capsid proteins [65]. Furthermore, coliphages inactivation was found to affected by pH and dissolved oxygen, thereby supporting the hypothesis stating that slower iron oxidation enhances virus inactivation. The adsorption of intermediates formed during the oxidation on the virus capsids was also necessary to inactivate the virus.

Heffron et al. suggested that coupling iron oxidation with regulated pH, similar to enhanced coagulation, will be able to achieve higher disinfection because of slow ferrous oxidation [66]. f2 and Qβ were also found to be removed at a pH range of 6–8 by ferrate oxidation [9, 13]. Also, it was found that ferrate action is not as influenced by the disinfecting matrix compared to that of chlorine [67].

5.3 Effect of ferrate ions on algae

Eutrophication of lakes and rivers has increased rapidly due to discharge of the nutrients into the water bodies, specifically phosphate. Thus, removal of these algal blooms is now a major challenge in the drinking water treatment. Conventional coagulation is presently the method of choice for algal removal in the treatment processes. Copper sulphate and potassium permanganate are chemicals that are most commonly employed to control the algal blooms in lakes and rivers. Studies suggest that pretreatment with an oxidant enhances the algal removal on coagulation. Ferrate being an excellent oxidant was thus checked for algal removal. Algal removal on the addition of alum combined with ferrate pretreatment was found to improve the coagulation of algal cells [68].

It was observed that ferrate peroxidation enhanced the coagulation of algae by hampering the stability of algal colloids. Ferrate oxidation results in the formation of $Fe(OH)_3$. $Fe(OH)_3$ precipitates on algal cell surface resulting in increased weight and thus enhanced settleablity. It was also hypothesized that on ferrate oxidation, the algal cells release certain exopolymer, which aids the coagulation. In a study conducted on *Microcystis aeruginosa,* it was observed that ferrate oxidation causes complete cell damage. Here the ferrate oxidation resulted in blocking photosynthesis by hampering the PSII functioning. The accumulation of excessive ROS further results in superoxidation of lipids, thereby increasing the cell permeability which finally results in complete cell damage of *Microcystis aeruginosa* cells. The removal of algae on ferrate was also influenced by pH and temperature. The removal efficiency was higher as the temperature increased, and also acidic pH enhanced the ferrate action on green algae [1, 10, 20].

5.4 Effect of ferrate ions on microcystins

Cyanobacteria such as *Microcystis*, *Anabaena*, *Aphanizomenon*, *Oscillatoria*, *Cylindrospermopsis* and *Lyngbya* are known to produce toxins, which affect aquatic life as well as human beings adversely. Some common toxins produced by cyanobacteria include anatoxin-a, anatoxin-as, aplysiatoxin, cylindrospermopsin, microcystin-LR, nodularin R and saxitoxin. Because of its adverse effects, US Environmental Protection Agency (EPA) in the year 2015 prescribed drinking water health advisories for microcystins and cylindrospermopsin. 0.3 mg L^{-1} was the prescribed limit for microcystins for preschool children, while the limit was 1.6 mg L^{-1} for school-age children.

The microcystins are composed of cyclic peptide ring of 7 amino acids and hepatotoxins, which are liver toxins. The peptide ring is composed of 5 non-protein amino acids and 2 protein amino acids. The microcystins' name is derived from these two protein amino acids. The most potent microcystin toxin is composed of leucine and arginine, thus the name microcystin-LR. Microcystin can remain in the water bodies from days to weeks, making it more dangerous. The toxicity of the microcystins is because of its Adda group [9, 14]. Ferrate essentially degrades the toxins by oxidation of the amino groups of microcystins. Thus, ferrate-induced degradation of algal toxins is greatly influenced by pH [69].

6. Conclusions

Ferrate [Fe(VI)] is a powerful oxidizing agent and can be used to remove organic and inorganic pollutants over a wide range of pH. Electrochemical synthesis of ferrate is the greenest and easiest method for high purity ferrate. The application of Fe(VI) for the removal of various recalcitrant pollutants such as pharmaceuticals, textile dyes, phenolic compounds, pesticides and inorganics, including heavy metals, is one of the promising technologies because it can act as oxidant as well as coagulant at the same time. The ability of ferrate to be an oxidant as well as coagulant also enhances removal of algae and microcystins. Also, ferrate is a potent disinfectant capable of removing bacteria as well as viruses. The action of ferrate, however, is dependent on pH and dissolved oxygen level of the media to be disinfected. The pH of the solution is one of the important factors which determines the stability and oxidation potential of Fe(VI). At higher pH values, the stability of Fe(VI) is higher as it increases the reaction rate with protonation of compounds. However, a high alkaline pH of liquid Fe(VI) causes the formation of harmful disinfection byproducts during disinfection treatment. Hence, a detailed research should be carried out for understanding the effect of pH in the treatment of various recalcitrant pollutants with different Fe(VI) concentrations.

Conflict of interest

The authors declare no conflict of interest.

Author details

Ansaf V. Karim, Sukanya Krishnan, Lakshmi Pisharody and Milan Malhotra*
Indian Institute of Technology, Mumbai, India

*Address all correspondence to: milanmalhotraiitb@gmail.com

IntechOpen

© 2020 The Author(s). Licensee IntechOpen. Distributed under the terms of the Creative Commons Attribution - NonCommercial 4.0 License (https://creativecommons.org/licenses/by-nc/4.0/), which permits use, distribution and reproduction for non-commercial purposes, provided the original is properly cited.

References

[1] Malhotra M, Suresh S, Garg A. Tea waste derived activated carbon for the adsorption of sodium diclofenac from wastewater: Adsorbent characteristics, adsorption isotherms, kinetics, and thermodynamics. Environmental Science and Pollution Research. 2018;**25** (32):32210-32220. DOI: 10.1007/s11356-018-3148-y

[2] Ansaf KVK, Ambika S, Nambi IM. Performance enhancement of zero valent iron based systems using depassivators: Optimization and kinetic mechanisms. Water Research. 2016;**102**:436-444. DOI: 10.1016/j.watres.2016.06.064

[3] Malhotra M, Garg A. Performance of non-catalytic thermal hydrolysis and wet oxidation for sewage sludge degradation under moderate operating conditions. Journal of Environmental Management. 2019;**238**:72-83. DOI: 10.1016/j.jenvman.2019.02.094

[4] Divyapriya G, Nambi IM, Senthilnathan J. Nanocatalysts in Fenton based advanced oxidation process for water and wastewater treatment. Journal of Bionanoscience. 2016;**10**: 356-368. DOI: 10.1166/jbns.2016.1387

[5] Tiwari D, Kim HU, Choi BJ, Lee SM, Kwon OH, Choi KM, et al. Ferrate(VI): A green chemical for the oxidation of cyanide in aqueous/waste solutions. Journal of Environmental Science and Health, Part A: Toxic/Hazardous Substances and Environmental Engineering. 2007;**42**:803-810. DOI: 10.1080/10934520701304674

[6] Graham N, Jiang CC, Li XZ, Jiang JQ, Ma J. The influence of pH on the degradation of phenol and chlorophenols by potassium ferrate. Chemosphere. 2004;**56**:949-956. DOI: 10.1016/j.chemosphere.2004.04.060

[7] Schreyer JM, Ockerman LT. Stability of ferrate(VI) ion in aqueous solution. Analytical Chemistry. 1951;**23**: 1312-1314. DOI: 10.1021/ac60057a028

[8] Licht S, Wang B, Ghosh S. Energetic iron (VI) Chemistry: The super-iron battery, Science. 1999;**285**:1039-1043

[9] Wagner WF, Gump JR, Hart EN. Factors affecting the stability of aqueous potassium ferrate(VI) solutions. Analytical Chemistry. 1952;**24**: 1497-1498. DOI: 10.1021/ac60069a037

[10] Virender S. K, potassium ferrate (VI): An environmentally friendly oxidant. Advances in Environmental Research. 2002;**6**:143-156. Available from: http://www.sciencedirect.com/science/article/pii/S1093019101001198

[11] Sharma VK, Kazama F, Hu J, Ray AK. Ferrates (iron(VI) and iron (V)): Environmentally friendly oxidants and disinfectants. Journal of Water and Health. 2005;**3**:45-58

[12] Jiang J, Wang S, Panagoulopoulos A. The exploration of potassium ferrate (VI) as a disinfectant/coagulant in water and wastewater treatment. Chemosphere. 2006;**63**:212-219. DOI: 10.1016/j.chemosphere.2005.08.020

[13] Williams DH, Riley JT. Preparation and alcohol oxidation studies of the ferrate (VI) ion, FeO_4^{2-}. Inorganica Chimica Acta. 1974;**8**:177-183. DOI: 10.1016/S0020-1693(00)92612-4

[14] Thompson GW, Ockerman L, Schreyer JM. Preparation and purification of potassium ferrate. VI. Journal of the American Chemical Society. 1951;**73**(3):1379-1381. DOI: 10.1021/ja01147a536

[15] Kopelev N, Perfiliev Y, Kiselev Y. Mössbauer study of sodium ferrates (IV) and (VI). Journal of Radioanalytical and Nuclear Chemistry.

1992;**162**(2):239-251. DOI: 10.1016/j.watres.2019.04.045

[16] Jiang J-Q, Lloyd B. Development of ferrate(VI) salt as an oxidant and coagulant for water and wastewater treatment. Water Research. 2002;**36**: 1397-1408. DOI: 10.4028/www.scientific.net/AMM.361-363.658

[17] Perfiliev YD, Tambiev AK, Konnychev MA, Skalny AV, Lobakova ES, Kirpichnikov MP. Mössbauer spectroscopic study of transformations of iron species by the cyanobacterium Arthrospira platensis (formerly Spirulina platensis). Journal of Trace Elements in Medicine and Biology. 2018;**48**:105-110. DOI: 10.1016/j.jtemb.2018.02.030

[18] Alsheyab M, Jiang JQ, Stanford C. On-line production of ferrate with an electrochemical method and its potential application for wastewater treatment—A review. Journal of Environmental Management. 2009;**90**: 1350-1356. DOI: 10.1016/j.jenvman.2008.10.001

[19] Sharma VK. Oxidation of inorganic contaminants by ferrates (VI, V, and IV)-kinetics and mechanisms: A review. Journal of Environmental Management. 2011;**92**:1051-1073. DOI: 10.1016/j.jenvman.2010.11.026

[20] Jiang JQ, Lloyd B. Progress in the development and use of ferrate(VI) salt as an oxidant and coagulant for water and wastewater treatment. Water Research. 2002;**36**:1397-1408. DOI: 10.1016/S0043-1354(01)00358-X

[21] Zhu JH, Yan XL, Liu Y, Zhang B. Improving alachlor biodegradability by ferrate oxidation. Journal of Hazardous Materials. 2006;**135**:94-99. DOI: 10.1016/j.jhazmat.2005.11.028

[22] Chen J, Qi Y, Pan X, Wu N, Zuo J, Li C, et al. Mechanistic insights into the reactivity of ferrate(VI) with phenolic compounds and the formation of coupling products. Water Research. 2019;**158**:338-349. DOI: 10.1016/j.watres.2019.04.045

[23] Chen J, Xu X, Zeng X, Feng M, Qu R, Wang Z, et al. Ferrate (VI) oxidation of polychlorinated diphenyl sulfides: Kinetics, degradation, and oxidized products. Water Research. 2018;**143**:1-9. DOI: 10.1016/j.watres.2018.06.023

[24] Yates BJ, Zboril R, Sharma VK. Engineering aspects of ferrate in water and wastewater treatment—A review. Toxic/Hazardous Substances and Environmental Engineering. 2014;**4529**. DOI: 10.1080/10934529.2014.950924

[25] Zhang P, Zhang G, Dong J, Fan M, Zeng G. Bisphenol A oxidative removal by ferrate (Fe(VI)) under a weak acidic condition. Separation and Purification Technology. 2012;**84**:46-51. DOI: 10.1016/j.seppur.2011.06.022

[26] Sun S, Jiang J, Qiu L, Pang S, Li J, Liu C, et al. Activation of ferrate by carbon nanotube for enhanced degradation of bromophenols: Kinetics, products, and involvement of Fe(V)/Fe(IV). Water Research. 2019; **156**:1-8. DOI: 10.1016/j.watres.2019.02.057

[27] Li G, Wang N, Liu B, Zhang X. Decolorization of azo dye Orange II by ferrate(VI)-hypochlorite liquid mixture, potassium ferrate(VI) and potassium permanganate. Desalination. 2009;**249**: 936-941. DOI: 10.1016/j.desal.2009.06.065

[28] Xu GR, Zhang YP, Li GB. Degradation of azo dye active brilliant red X-3B by composite ferrate solution. Journal of Hazardous Materials. 2009; **161**:1299-1305. DOI: 10.1016/j.jhazmat.2008.04.090

[29] Turkay O, Barışçı S, Dimoglo A. Kinetics and mechanism of methylene

blue removal by electrosynthesized ferrate(VI). Separation Science and Technology. 2016;**51**:1924-1931. DOI: 10.1080/01496395.2016.1182189

[30] Wu S, Li H, Li X, He H, Yang C. Performances and mechanisms of efficient degradation of atrazine using peroxymonosulfate and ferrate as oxidants. Chemical Engineering Journal. 2018;**353**:533-541. DOI: 10.1016/j.cej.2018.06.133

[31] Chen Y, Xiong Y, Wang Z, Chen Y, Chen G, Liu Z. UV/ferrate(VI) oxidation of profenofos: Efficiency and mechanism. Desalination and Water Treatment. 2015;**55**:506-513. DOI: 10.1080/19443994.2014.917987

[32] Liu H, Pan X, Chen J, Qi Y, Qu R, Wang Z. Kinetics and mechanism of the oxidative degradation of parathion by ferrate(VI). Chemical Engineering Journal. 2019;**365**:142-152. DOI: 10.1016/j.cej.2019.02.040

[33] Liu H, Chen J, Wu N, Xu X, Qi Y, Jiang L, et al. Oxidative degradation of chlorpyrifos using ferrate(VI): Kinetics and reaction mechanism. Ecotoxicology and Environmental Safety. 2019;**170**: 259-266. DOI: 10.1016/j.ecoenv.2018.11.132

[34] Sharma VK, Mishra SK, Ray AK. Kinetic assessment of the potassium ferrate(VI) oxidation of antibacterial drug sulfamethoxazole. Chemosphere. 2006;**62**:128-134. DOI: 10.1016/j.chemosphere.2005.03.095

[35] Sharma VK, Mishra SK. Ferrate(VI) oxidation of ibuprofen: A kinetic study. Environmental Chemistry Letters. 2006;**3**:182-185. DOI: 10.1007/s10311-005-0002-5

[36] Ma Y, Gao N, Li C. Degradation and pathway of tetracycline hydrochloride in aqueous solution by potassium ferrate. Environmental Engineering Science. 2012;**29**:357-362. DOI: 10.1089/ees.2010.0475

[37] Wang Y, Liu H, Liu G, Xie Y, Gao S. Oxidation of diclofenac by potassium ferrate (VI): Reaction kinetics and toxicity evaluation. Science of the Total Environment. 2015;**506–507**:252-258. DOI: 10.1016/j.scitotenv.2014.10.114

[38] Jiang JQ, Zhou Z, Patibandla S, Shu X. Pharmaceutical removal from wastewater by ferrate(VI) and preliminary effluent toxicity assessments by the zebrafish embryo model. Microchemical Journal. 2013;**110**:239-245. DOI: 10.1016/j.microc.2013.04.002

[39] Wang S, Hu Y, Wang J. Strategy of combining radiation with ferrate oxidation for enhancing the degradation and mineralization of carbamazepine. Science of the Total Environment. 2019;**687**:1028-1033. DOI: 10.1016/j.scitotenv.2019.06.189

[40] Barişçi S, Ulu F, Sillanpää M, Dimoglo A. The usage of different forms of ferrate (VI) ion for amoxicillin and ciprofloxacin removal: Density functional theory based modelling of redox decomposition. Journal of Chemical Technology and Biotechnology. 2016;**91**:257-266. DOI: 10.1002/jctb.4625

[41] Yang B, Ying GG, Zhao JL, Liu S, Zhou LJ, Chen F. Removal of selected endocrine disrupting chemicals (EDCs) and pharmaceuticals and personal care products (PPCPs) during ferrate(VI) treatment of secondary wastewater effluents. Water Research. 2012;**46**: 2194-2204. DOI: 10.1016/j.watres.2012.01.047

[42] Zhou Z, Jiang JQ. Treatment of selected pharmaceuticals by ferrate(VI): Performance, kinetic studies and identification of oxidation products. Journal of Pharmaceutical and

Biomedical Analysis. 2015;**106**:37-45. DOI: 10.1016/j.jpba.2014.06.032

[43] Johnson MD, Hornstein BJ. The kinetics and mechanism of the ferrate (VI) oxidation of hydroxylamines. Inorganic Chemistry. 2003;**42**: 6923-6928. DOI: 10.1021/ic020705x

[44] Lee Y, Zu C. Ferrate (Fe(VI)) application for municipal wastewater treatment: A novel process for simultaneous micropollutant oxidation and phosphate removal. Environmental Science & Technology. 2009:3831-3838

[45] Lee Y, Um I. Arsenic (III) oxidation by iron (VI) (ferrate) and subsequent removal of arsenic (V) by iron (III) coagulation. Environmental Science & Technology. 2003;**37**:5750-5756

[46] Lee Y, Um I, Yoon J. Ferrate (VI) oxidation of arsenite (III) and subsequent removal of arsenate (V) by iron (III) coagulation. Environmental Science & Technology. 2003;**37**:3-5

[47] Lan B, Wang Y, Wang X, Zhou X, Kang Y, Li L. Aqueous arsenic (As) and antimony (Sb) removal by potassium ferrate. Chemical Engineering Journal. 2016;**292**:389-397. DOI: 10.1016/j.cej.2016.02.019

[48] Sylvester P, Rutherford LA, Gonzalez-Martin A, Kim J, Rapko BM, Lumetta GJ. Ferrate treatment for removing chromium from high-level radioactive tank waste. Environmental Science & Technology. 2001;**35**:216-221. DOI: 10.1021/es001340n

[49] Tiwari D, Sailo L, Pachuau L. Remediation of aquatic environment contaminated with the iminodiacetic acid metal complexes using ferrate(VI). Separation and Purification Technology. 2014;**132**:77-83. DOI: 10.1016/j.seppur.2014.05.010

[50] Sailo L, Pachuau L, Yang JK, Lee SM, Tiwari D. Efficient use of ferrate(VI) for the remediation of wastewater contaminated with metal complexes. Environmental Engineering Research. 2015;**20**:89-97. DOI: 10.4491/eer.2014.079

[51] Yngard R, Damrongsiri S, Osathaphan K, Sharma VK. Ferrate(VI) oxidation of zinc-cyanide complex. Chemosphere. 2007;**69**:729-735. DOI: 10.1016/j.chemosphere.2007.05.017

[52] Lee SM, Tiwari D. Application of ferrate(VI) in the treatment of industrial wastes containing metal-complexed cyanides: A green treatment. Journal of Environmental Sciences. 2009;**21**:1347-1352. DOI: 10.1016/S1001-0742(08)62425-0

[53] Osathaphan K, Tiyanont P, Yngard RA, Sharma VK. Removal of cyanide in Ni(II)-cyanide, Ni (II)-cyanide-EDTA, and electroplating rinse wastewater by ferrate(VI). Water, Air, and Soil Pollution. 2011;**219**: 527-534. DOI: 10.1007/s11270-010-0725-1

[54] Pachuau L, Lee SM, Tiwari D. Ferrate(VI) in wastewater treatment contaminated with metal(II)-iminodiacetic acid complexed species. Chemical Engineering Journal. 2013; **230**:141-148. DOI: 10.1016/j.cej.2013.06.081

[55] Jiang JQ, Panagoulopoulos A, Bauer M, Pearce P. The application of potassium ferrate for sewage treatment. Journal of Environmental Management. 2006;**79**:215-220. DOI: 10.1016/j.jenvman.2005.06.009

[56] Rai PK, Lee J, Kailasa SK, Kwon EE, Tsang YF, Ok YS, et al. A critical review of ferrate(VI)-based remediation of soil and groundwater. Environmental Research. 2018;**160**:420-448. DOI: 10.1016/j.envres.2017.10.016

[57] Lee Y, Zimmermann SG, Kieu AT, Von Gunten U. Ferrate (Fe(VI)) application for municipal wastewater treatment: A novel process for simultaneous micropollutant oxidation and phosphate removal. Environmental Science & Technology. 2009;**43**: 3831-3838. DOI: 10.1021/es803588k

[58] Lim M, Kim MJ. Removal of natural organic matter from river water using potassium ferrate(VI). Water, Air, and Soil Pollution. 2009;**200**:181-189. DOI: 10.1007/s11270-008-9902-x

[59] Gombos E, Barkács K, Felföldi T, Vértes C, Makó M, Palkó G, et al. Removal of organic matters in wastewater treatment by ferrate (VI)-technology. Microchemical Journal. 2013;**107**:115-120. DOI: 10.1016/j.microc.2012.05.019

[60] Jiang JQ. Research progress in the use of ferrate(VI) for the environmental remediation. Journal of Hazardous Materials. 2007;**146**:617-623. DOI: 10.1016/j.jhazmat.2007.04.075

[61] Nguema PF, Jun M. Application of ferrate (VI) as disinfectant in drinking water treatment processes: A review. International Journal of Microbiology Research. 2016;**7**:53-62. DOI: 10.5829/idosi.ijmr.2016.53.62

[62] Gilbert MB, Waite TD, Hare C. Analytical notes — an investigation of the applicability of ferrate ion for disinfection. Journal of American Water Works Association. 1976;**68**(9):495-497

[63] Hjelmsø MH, Hellmér M, Fernandez-Cassi X, Timoneda N, Lukjancenko O, Seidel M, et al. Evaluation of methods for the concentration and extraction of viruses from sewage in the context of metagenomic sequencing. PLoS One. 2017;**12**:e0170199. DOI: 10.1371/journal.pone.0170199

[64] Kazama F. Respiratory inhibition of Sphaerotilus by potassium ferrate. Journal of Fermentation and Bioengineering. 1989;**67**:369-373. DOI: 10.1016/0922-338X(89)90042-1

[65] Hu L, Page MA, Sigstam T, Kohn T, Mariñas BJ, Strathmann TJ. Inactivation of bacteriophage MS2 with potassium ferrate(VI). Environmental Science & Technology. 2012;**46**:12079-12087. DOI: 10.1021/es3031962

[66] Heffron J, McDermid B, Mayer BK. Bacteriophage inactivation as a function of ferrous iron oxidation. Environmental Science: Water Research & Technology. 2019;**5**: 1309-1317. DOI: 10.1039/c9ew00190e

[67] Schink T, Waite TD. Inactivation of f2 virus with ferrate (VI). Water Research. 1980;**14**:1705-1717. DOI: 10.1016/0043-1354(80)90106-2

[68] Ma J, Liu W. Effectiveness and mechanism of potassium ferrate(VI) preoxidation for algae removal by coagulation. Water Research. 2002;**36**: 871-878. DOI: 10.1016/S0043-1354(01)00282-2

[69] Deng Y, Wu M, Zhang H, Zheng L, Acosta Y, Hsu TTD. Addressing harmful algal blooms (HABs) impacts with ferrate(VI): Simultaneous removal of algal cells and toxins for drinking water treatment. Chemosphere. 2017;**186**: 757-761. DOI: 10.1016/j.chemosphere.2017.08.052

Section 4

Industrial Applications Trends and Prospects

Chapter 6

Application of Advanced Oxidation Process in the Food Industry

Zhaoran Xin and Lars Rehmann

Abstract

Wastewater in the food industry contains recalcitrant organic compounds and a certain degree of toxicity. Present wastewater treatment plants are insufficient in dealing with the increasing complexity of effluents from modern food industries. Improperly treaded wastewaters can lead to spoil soil and are threads to aquatic life. The reaction of these recalcitrant chemicals with reactive radicals is an efficient treatment strategy. Researchers have proposed advanced oxidation processes (AOPs) that generate reactive radicals including ozonation, UV irradiation, (photo-) Fenton process, etc. This chapter reviews laboratory-scale and pilot-scale AOPs to incorporate with conventional pre-treatment methods and to evaluate their effectiveness and factors including operation condition and catalysts to optimize the process. Further research related to novel catalyst synthesis and cost evaluation of pilot-scale study is suggested.

Keywords: advanced oxidation processes (AOPs), food industry wastewaters, physicochemical methods, biodegradability improvement, combined treatment

1. Introduction

Water is a widely used resource in the food industry such as sanitary water for food processing, as a raw material, and for equipment cleaning. The large consumption of water leads to a corresponding generation of wastewater. Wastewater treatment from a diverse industry is technically more challenging than the treatment of domestic wastewater whose characteristics are largely similar. The various utilizations of water in the industries result in complex compositions of the resulting wastewater [1], which is often a thread to human, plant, and aquatic life [2].

The complexity of wastewaters from the food industry is a more severe issue than its amount. The effluents typically include large total load of organic pollutants such as colored and recalcitrant compounds with high organic loads (chemical oxygen demand, COD), proteins or fat, and chemicals used during processing [1], some with considerable levels of toxicity [3]. Current municipal wastewater treatment plants with the remaining capacity for population growth are not designed to handle the characteristics of such influent loading [4], creating a demand of complementary technologies in addition to traditional treatments.

Advanced oxidation processes (AOPs) generate highly reactive radical (**Table 1**). As one of the most active oxidizing agents, hydroxyl radical is generally

IntechOpen

Oxidizing agent	Relative oxidation activity
Positively charged hole on titanium dioxide, TiO_2^+	2.35
Hydroxyl radical	2.05
Atomic oxygen	1.78
Ozone	1.52
Hydrogen peroxide	1.31
Permanganate	1.24
Hypochlorous acid	1.1
Chlorine	1

Table 1.
Relative oxidation activity of some oxidizing agents [5].

Organic compound	Rate constant $[M^{-1} s^{-1}]$	
	O_3	HO
Alcohols	10^{-2}–1	10^8–10^9
Aromatics	1–10^2	10^8–10^{10}
Chlorinated alkenes	10^3–10^4	10^9–10^{11}
Ketones	1	10^9–10^{10}
N-containing organics	10–10^2	10^8–10^{10}
Phenols	10^3	10^9–10^{10}

Table 2.
Reaction rate constants for ozone and hydroxyl radical for organic compounds [7].

used in AOPs. It reacts nonselectively with organic compounds [6] which are resistant to conventional oxidation methods as high reaction rate (**Table 2**).

AOPs are widely used in winery wastewater, olive oil mill wastewater, and slaughterhouse wastewater. In addition to degradation of extra chemical compounds directly, AOPs are used to degrade cellular contents to disinfect wastewater with high organic nature in the dairy industry [8].

However, AOPs are capital-intensive since they include the use of expensive regents such as ozone and hydrogen peroxide, as well as the cost of equipment including sources of ultraviolet light. Hence AOPs are often combined with traditional treatments to overcome this drawback. The combination technologies reduce toxicity and increase the biodegradability of wastewater effluent after AOPs, which is more suitable for inexpensive biological process.

As a complementary of previous reviews [1, 9], this review categorizes relevant AOP applications in the food sector into winery, olive oil mill, meat industries, dairy, and food dye sections. This work summarizes state-of-the-art wastewater treatment technologies in food industries by AOPs both at laboratory-scale and pilot-scale after 2016. Current treatment technologies are also introduced.

2. Introduction to current wastewater treatment technologies in the food industry

Current wastewater treatment in the food industry is evolving from traditional treatments by focusing on the degradation of excessive organic matters, the recovery of profitable by-products, and water reusability. Typically wastewater treatment consists of three common stages: primary treatment such as physicochemical treatments partially removing solids; secondary treatment aimed at removing biodegradable organics and suspended solids using biological methods including aerobic and anaerobic processes; and tertiary treatments designed to remove residual compounds of particular concern to the given plant. Current technology alternatives to AOPs for wastewater treatment in the food industry are introduced briefly.

2.1 Evaporation

Natural evaporation during summer can be applied to olive oil mill wastewater (OMW) stored in lagoons, which reduces the volume of waste but results in odor and soil pollution. Alternatively, vacuum evaporation can reduce the volume of wastewater at low temperature and pressure; two streams including distillate and concentrate are separated which can then easily be handled with common treatments [10].

2.2 Land application

The application of wastewater as organic fertilizers has been investigated. High organic loadings in wastewater provide nutrients for soil but can also result in soil contamination. For lower amounts of wastewater, soil can act as biofilter of OMW degradation [11]. In contrary to direct application, composting is more applicable due to the elimination of phenols [12].

2.3 Physicochemical treatments

Colloidal particles in wastewater result in filtering devices clog. To prevent the malfunction of equipment for biological treatment and to meet the discharge requirement, technologies called physicochemical treatments are in demand to eliminate these particles.

The addition of flocculation destabilizes colloidal dispersions which count for the majority of the total dispersed solids (TDS), turbidity, and part of COD. Moreover, lime treatment and adsorption are also practical technologies [13]. Such treatments increase biodegradability during biological treatment and reduce residual components including organic load, color, and metal content after biological treatment [14]. For further study, alternative coagulants are promising to achieve higher removal efficiency, less pollution, and lower health risks [15].

2.4 Membrane technologies

Membrane technology is applied when phytotoxic recalcitrant pollutants invalidate biological treatment [16] or for recovery of valuable products [17]. With the advantage of high efficiency, simple equipment, and convenient operation, membrane technologies are beneficial to small plants that cannot afford high investments. However, membrane fouling needs to be reduced to increase economic feasibility. Pulido et al. reviewed current membrane technologies addressing the efforts of decreasing fouling issues in OMW [16].

According to the comprehensive review papers of winery wastewater treatment, membrane bioreactors are competitive considering water reuse [14, 15].

2.5 Summary

The choice of treatment solution is based on the specific characteristic of wastewater. The combination of complementary treatments is promising to achieve discharge regulation. More detailed current technologies and their comparison are reviewed elsewhere [18].

The advances in wastewater treatment in the food industry are driven by the rising awareness of environment protection in the public. At present, for example, various reactors are being designed for olive mill wastewater treatment, while most of them were not treated before 2004 in the EU [10]. However, the remaining challenges after current treatment call for extra treatment technologies in future research, which might be addressed through AOPs.

3. Application of AOPs in the food industry

3.1 Wineries

In 2016, the worldwide total alcohol consumption was equal to 6.4 liters of pure alcohol per person aged 15 years and older [19]. Accordingly, winery wastewater originates from multiple steps in the production process. One of the origins is fresh water used for cleaning, including washing of equipment and facilities [20], while other streams are directly related to wine, such as the effluent from filtration units and off products [15].

Winery wastewater contains organic matters such as ethanol, sugars, organic acids, phenolic compounds, etc. [15]. Moreover, there are toxic compounds from pesticides used on grapes [22]. Winery wastewater discharged without any treatment would result in soil pollution. General sewage treatment facilities are not sufficient to process the wastewater which leads to acid pH and high COD [27]. The treatment for winery wastewater has to be specific due to the differences between each winery, such as the type of wine, unique component and volume of the wastewater, etc. [28]. Even in the same winery, the effluent is unstable according to the operation period and season [29].

A recent review [15] summarizes the processes currently applied and/or tested for the treatment including physicochemical, biological, membrane filtration and separation, advanced oxidation processes, and combined biological and advanced oxidation processes. Physicochemical processes (i.e., coagulation/flocculation, EC) can be used as pre-treatment to lower TSS. Biological treatment processes using membrane bioreactors (MBR) are promising and are efficient in reducing organic load, while AOPs, especially photo-Fenton as post-treatment in combination with other treatment, show higher efficiency in reducing COD. Mechanism and restriction of each method are reviewed elsewhere [30], also providing a guidance for choosing suitable method according to different matrices. Applications of AOPs in the winery industry since 2016 are summarized in **Table 3**.

The combined electro-Fenton processes and photolysis processes showed higher efficiency in degradation rate and energy consumption because of synergetic effect. Such processes can be implemented in separate steps or can be integrated together. Díez et al. proposed different sequential reactors which are divided into two sections. Wastewater processed in the electro-Fenton section generates Fe complexes, which are decomposed by light radiation in the photolysis section. They initially proposed a two-chamber cubic reactor, which suffered from foam formation during aeration. The foam reduces the irradiation volume; hence the reactor was changed

Winery wastewater origin	Treatment process	Observations	Reference
Simulated winery wastewater (SWW) from commercial wine. COD 14.43 g/L, TOC 3.726 g/L, maximum wavelength 522 nm, color intensity (CI) 1.31, browning index (BI) 0.55, pH 3.9	Photo-electro-Fenton process	TOC reduced to 61% (mercury lamp) and 65% (LED lamp). DC, CI, and BI reduced totally. (optimized condition: electrode gap, 2 cm electric field 15 V; reaction time, 180 min) TOC: 68.14% (RWW1), 68.77% (RWW2)	[21]
Simulated (SE1, SE2) and real (RE) COD 10,940 mg/L TOC 4427 mg/L pH 3.2 COD; TOC of SE2 is about 8 times of SE1 (82,050, 33,200). pH 2.9 COD (86100) of real WW (60100) is similar to SE2; TOC is twofolds of SE2. pH 3.4	Photo-electro-Fenton process	SE1(200 min) Color: 97.49% TOC: 53.01% G-PTFE cathode, flow rate: 2 ml/min SE2: Color: 93% TOC: 65.7% RE: Color: 60% TOC: 51% 5 V, reaction time 1.3 h	[22]
local white wine producer (Spain) pH 4.08 TOC 55 g L^{-1}	Sequential two-column electro-Fenton-photolytic reactor	Color: 65% TOC: 67%, 64%^{P} COD: 77%, 74%^{P} IC_{50}: 76.5%, 41%^{P} M Na_2SO_4, 75 mg L^{-1} of Fe^{3+}, pH 2 and 5 V Reaction time: 12 h^P: with pesticides	[22]
Undurraga® Winery Company (Talagante, Santiago de Chile) COD 3490 mg/L TOC 1320 mg/L NTU 15.2	Anodic oxidation	Eliminate COD, TOC, and NTU totally 50 mM of NaCl or Na_2SO_4 and apply higher-density currents (60 mA /cm^2) Reaction time 420 min pH 8.30 ± 0.20	[23]
Douro red wine diluted samples COD 513 mg/L TOC 143 mg/L pH 4.0	Solar radiation-assisted sulfate radical-based advanced oxidation processes (SR-AOP)	TOC: around 50% COD: 75% (pH 4.5, PMS, Fe (II)) Reaction time 3 h	[24]
Wine cellar in the Douro region in the north of Portugal pH = 4.37, COD 600 mg/L, BOD_5 145 mg/L, BOD_5/COD 0.24. TOC 166 mg/L	UV-C assisted sulfate radical-based advanced oxidation processes (SR-AOP)	COD 96%, TOC 71% Reaction time 240 min. pH 7. Consumed 25 mM $S_2O_8^{2-}$ UV-C = 254 nm	[25]
pH = 3.8, COD 2.128 g/L, BOD_5 974 mg/L, TOC 825 mg/L	Adsorption and photo-Fenton process Ca-smectite as adsorbent and catalyst support	TOC reduced 90% in total, by 54% (adsorption) and 36% (photo-Fenton) pH = 4, H_2O_2 98 MM, catalyst 6 g/L	[26]

Table 3.
Recent application of AOPs in the winery industry.

into two divided columns. The economic efficiency was increased by decreasing the voltage from 15 to 5 V [21, 22]. In order to minimize the electrode gap, the reactor system is designed in two columns connected vertically. The efficiency for degradation of TOC and decreasing color increased in treating real winery wastewater compared to initial laboratory studies.

Both studies investigating simulated winery wastewater using column reactors eliminate color totally. Other than color, TOC reduction rates are both nearly 65%. However, the efficiency reduced with an increased concentration of TOC, which reduces ·OH from UV radiation, and accumulated carboxylic acids inhibit regeneration of Fe(II) by reacting with Fe(III) [22].

Anodic oxidation with boron-doped diamond (BDD) electrode is able to process real effluents with high concentration of COD (3490 mg/L), TOC (1320 mg/L), and more than 40 organic compounds [23]. The addition of electrolytes including NaCl and Na_2SO_4 and the application of high-density currents (60 mA/cm^2) increased the efficiency of oxidation, leading to total mineralization in 400 min. The higher efficiency of this process results from the larger number of radicals which have two origins: (i) HO· radicals are weakly bounded on BDD electrode with high O_2-overpotential; (ii) addition of salt generates other radicals such as $S_2O_8{}^{2-}$ and active chlorine species.

Solar-driven sulfate radical is also a promising tertiary treatment [24]. This process has the advantage of independence from pH. The addition of transition metal as catalyst including Fe(II) and Co(II) shows that Fe(II) performs best when considering environmental implications and efficacy. Furthermore, the reduction of TOC could be enhanced from around 50–71% by combination of UV radiation and thermally activated persulfate (TAP) [25]. It indicates higher efficiency of $S_2O_8{}^{2-}$ than H_2O_2 for removal of COD (96%) and organic matters, as the UV-C/$S_2O_8{}^{2-}$ process generates two kinds of radicals, HO· and $SO_4{}^{-}$· ,where the yield of $SO_4{}^{-}$· is higher and $SO_4{}^{-}$· can be converted into HO·.

Natural Ca-smectite can be used for removal [26]. It works both as adsorbent and catalyst support saturated with Fenton. The adsorption capacity can be predicated by Jovanovich isothermal model. For photo-Fenton process' operation condition regarding pH, H_2O_2 concentration and catalyst dosage were optimized, yielding 90% of TOC removal with 54% due to adsorption and 36% due to the photo-Fenton process. However, catalyst regeneration analysis in three consecutive cycles shows reduction to 57%.

Based on the complex organic components and toxicity of winery wastewater, each method aims at increasing the effective radicals. Further study related to reactor optimization, utilization of multiple radicals, and novel catalyst synthesis is worthwhile.

3.2 Olive oil mills

The high demand of water in olive oil production processes results in large amounts of wastewater ranging from 0.5 to 1.68 m^3 per ton with large amounts of semisolid or slurry wastes.

Factors influencing the characteristics of olive oil mill wastewater (OMW) include (i) composition of vegetation water, (ii) olive oil extraction process, and (iii) storage time [30]. Moreover, phenolic compounds result in toxicity of the effluents. However, properly treated OMW produces nutrients (N, P, K) for plants [31]. Therefore, pre-treatment steps are necessary to economically increase the efficiency of the following treatment steps, as summarized in **Table 4**.

A photo-Fenton with medium pressure UV-lamp process was developed for mixture of real OMW [32]. This process showed the ability of fast degradation of pollutants (>90% in 30 mins). The wavelength of UV light did not influence the reduction rates. Another factor is the concentration of H_2O_2. The total nitrogen content removal rate is not related to the concentration of H_2O_2 but the reaction

Olive oil mill wastewater origin	Treatment process	Observations	Reference
Olive oil mill in the province of Seville (Spain) mixture of WOW (olives washing) and WOOW (olive oil washing) pH 5.98; COD 7060 mg/L; TOC 918 mg/L; BOD_5 685 mg/L; turbidity (FTU) 1390	Photo-Fenton	Nitrogen content removed 62.5–75.5%, COD = 95.7 ± 0.53%, TOC 96.3%, total phenolic compounds = 93.6 ± 2.5%, total carbon = 94.0 ± 1.2%, total organic carbon = 96.3 ± 0.6%, total nitrogen = 74.9 ± 6.8%, turbidity = 92.5 ± 1.9%. pH = 3, T = 20°C, catalyst = 3 g/L, reaction time 5–30 min, agitation rate = 600 rpm	[32]
Adjusted olive oil mill (mean value) samples from Cyprus, Israel, Jordan, and Portugal. pH 5.2 EC 12.5 mS/cm COD 25.0 g/L BOD_5 5.1 g/L DOC 6 g/L TSS 24 g/L TP 4.2 g/L TPh 4.2 g/L	Coagulation/flocculation followed by solar photo-Fenton oxidation	Coagulation/flocculation TSS 90%, COD 40%, DOC 11% Solar photo-Fenton oxidation COD 94%, DOC 43%, BOD_5 86%, TSS 96%, TPh 99.8% Reaction time 70 min	[17]
Badajoz (Spain) pH 4.9 COD 6450 mgO_2/L TN 42 mg/L TP 21 mg/L TSS 3190 mg/L BOD_5 2130 mg O_2/L BOD_5 /COD 0.33	Coagulation/flocculation followed by Fenton oxidation and biological treatment (only industry)	Lab Coagulation/flocculation COD 38% + 75% TSS 40% Industry COD 95% Flow rate 1.5m^3 h^{-1} Reaction time 60 days	[33]
COD 1344 ± 22 mg/L; TN 22.4 ± 1.4 mg/L; total phosphate 19.1 ± 1.6 mg/L; phenols 5.89 ± 0.8 mg/L; pH 6.9; conductivity 1260 ± 32 μS/cm Color, expressed as Pt/Co units 2310 ± 9 Olive Oil washing wastewater after filtration Mersin, Turkey	Combined electrocoagulation (ECR)-photocatalytic (PCR) degradation system	**ECR + PCR** COD 88% Phenol (ECR: 88%) 100% Color 100% Reaction time **PCR + ECR** COD 78% Phenol 93.2% (PCR: 40.7%) **ECR:** voltage, 12.5 V; CD, 12.9 A/m2; pH, 6.9; reaction time, 120 min; electrode type, Al **PCR:** catalyst loading, 1 g/L; catalyst type, ZnO; pH, 7.7; reaction time, 120 min; light source, (UVA)	[34]
COD 33927 ± 200 mg/L DOC 10120 mg/L Phenol 164 ± 13 mg/L Color 13,350 ± 120 (Pt-Co) Acid-cracked wastewater from olive production factory. Kahramanmaras, Turkey	Combined ozone/Fenton process	Color 51.6% +21% DOC 27.9% + 49% CODs 58% +22% Phenol 100%	[35]

Olive oil mill wastewater origin	Treatment process	Observations	Reference
TOC 23231–34,050 mgL^{-1} COD 62400–82,400 mgL^{-1} Color 369 Pt-Co pH 4.5 BOD_5/COD 0.144 100 times diluted sample wastewater from olive production factory. Gemlik region, Turkey	Microwave (MW) activated persulfate	TOC 100% COD 63.38% Color 94.85% Optical density 121.7% PS: 266 g L^{-1} Reaction time 23.58 min Power level 567 W Initial pH 2	[36]

Table 4.
Recent application of AOPs on OMW.

time. The removal efficiency decreased with longer reaction time since more N_2 is fixed with abundant CO_2. The conversion rates of TOC have all reached more than 90% and increased slightly with increasing H_2O_2 concentrations.

Solar photo-Fenton oxidation pretreated by coagulation/flocculation achieved 90% TSS removal at optimal dose of $FeSO_4{\cdot}7H_2O$ [17]. After coagulation, solar photo-Fenton oxidation removed up to 94% of COD. In this process, Fe^{2+} should be chosen to promote the regeneration of Fe(II). Otherwise, at higher Fe(II) concentration, the penetration efficiency of light decreases. At lower Fe(II) concentration, it accelerates consumption of H_2O_2 without the generation of ·OH, which also explained the low efficiency in high H_2O_2 concentration. Furthermore, the solar photo-Fenton-treated effluents contained fewer toxic matters and lead to higher nutrient uptake by plant. It appears to an efficient and economic technology for plants which can take advantage of sunshine. A similar technology was applied without solar photocatalysis, and the removal efficiency of COD decreased to 75% [33]. However, it increased to 95% after 60 days biological treatment since the process increases the biodegradability. Similar effects for different dosages of Fe^{2+} were observed. Scale-up experiment in industry indicates that adjusting the reagent dosages by monitoring the concentration of COD is necessary to optimize efficiency and cost. Compared with lab scale, decreased efficiency of degradation might be a result of the complexity of real matrices.

Optimal operation condition for electrocoagulation (ECR) and photocatalytic (PCR) process were investigated [34]. For ECR, neutral pH and higher current density are chosen to optimize the solubility of aluminum hydroxide, which enhances electrocoagulation of COD, phenol, and color. For PCR, different types of catalyst, including ZnO and TiO_2, had similar removal efficiency of COD and color. But ZnO has higher efficiency in phenol removal while also having a cost advantage. Sequential application of ECR before PCR results in higher degradation in COD and phenol.

Fenton process followed by ozonation has also been optimized [35]. Considering removal efficiency and cost of ozone, 90 min was found to be the optimal reaction time. However, subsequent biological treatment processes are needed to meet regulatory guidelines for COD and color.

Oxidation of OMW by microwave (MW)-activated persulfate is a promising pre- or post-treatment step [36]. Based on dielectric heating principle, MW irradiation is efficient in heating, which activates persulfate and breaks down phenolic compounds which have dark brown color. Hence MW counts for main removal of TOC, operating cost, and a portion of color reduction, while pH is the dominant factor in reducing color.

Rich in nutrient and color, more OMW is degraded by Fenton process in combination with pre-treatment technologies after 2016, compared with previous comprehensive literature review [13]. Factors involved in Fenton process are investigated extensively, including dosage of reactants and catalyst, pH, and reaction time. For further study, different combination of conventional pre-treatment methods and Fenton process can be analyzed. In addition, great effort for scale-up study and cost evaluation is needed.

3.3 Meat processing industry

The meat processing industry produces large volumes of slaughterhouse wastewater (SWW). SWW contains elevated amount of organic matter, suspended solids, oil, grease, and toxic matters [37]. Anaerobic treatment is efficient in removing organic matter with low costs in addition to the generation of methane. However, complementary treatment is necessary for the effluents to meet the required discharge limits [38]. As post-treatment method, AOPs are a promising supplementary for the entire process in removing non-biodegradable organics and inactivated microorganisms producing hazardous by-products [37] as summarized in **Table 5**.

UV/H_2O_2 has been applied to degrade slaughterhouse effluents pretreated by anaerobic baffled reactor (ABR) and an aerobic activated sludge (AS) reactor connected sequentially [37]. RSM was used to optimize the operation condition in order to maximize removal efficiency of TOC and TN and CH_4 yield, as well as minimize H_2O_2 in effluent.

This process can be enhanced by silver-doped TiO_2 nanoparticles as photocatalyst [42]. Electrons captured by silver react with oxygen more efficiently than those combined with electron hole in TiO_2. However, application of silver and further treatment for separation of photocatalyst make it an uneconomic technology for practical use.

The combined UV-C/H_2O_2-VUV was studied with the similar method used by Bustillo-Lecompte, but a Box–Behnken design was used instead of the CCD [39]. The combined process removed more TOC and minimized H_2O_2 simultaneously compared with individual processes. Furthermore, this technology performs better if applied after biological treatment, since the respirometry analyses indicated low biodegradable degree. Wastewater pretreated by anaerobic digestion can yield up to 90% COD removal but remains yellow in color after 30 days [41]. Further treatment is necessary to remove more COD and eliminate color compounds, which can be achieved by SPEF with BDD anode. SPEF results in higher decline of absorbance and COD than photolysis only and less cost compared with electron-Fenton (EF) [42] and Fenton process only [40].

Compared with conventional disinfection methods, cold plasma technology has the advantage of less by-product and high removal efficiency [43]. This technology achieved high removal efficiency of pathogens, organics, and inorganic pollutants at the same time with enough hydraulic retention time. Furthermore, other operation condition for cold plasma can be investigated.

Various AOPs combined with biological treatment methods have been applied. Treatment of wastewater from meat processing industry meets specific challenge including methane production. Different experiment strategies are performed to study the factors. Cost evaluation in scale-up plant is needed considering both removal efficiency and overall economy, including cost and profits of CH_4.

3.4 Dairy wastewater

The consumption of dairy production in Canada increases gradually. Correspondingly, more wastewaters need to be treated. Dairy industries' wastewaters come from processing milk and system management and typically contain large

Meat processing plants wastewater origin	Treatment process	Observations	Reference
BOD 1209 mg/L COD 4221 mg/L TN 427 mg/L TOC 546 mg/L TP 50 mg/L TSS 1164 mg/L pH 6.95 Slaughterhouse effluents in Ontario, Canada	An anaerobic baffled reactor (ABR), followed by an aerobic activated sludge (AS) reactor, and a UV/H_2O_2 photoreactor (ABR-AS-UV/H_2O_2 processes)	TOC 97.8% H_2O_2 residual 1.3% TOC 50 mg/L, flow rate 15 mL/min, H_2O_2 dosage 344 mg/L, pH 7.2	[37]
Slaughterhouse effluents in Ontario, Canada	Combined UV-C/H_2O_2-VUV	TOC 46.19% H_2O_2 residual 1.05% TOC 213 mg/L H_2O_2 dosage 450 mg/L Irradiation time 9 min	[39]
pH 7–8.3 COD 25–32 g/L Volatile solid (VS) 3.31–9 g/L TS 5.32–11.5 g/L Total coliforms MPN/100 mL 90x105 Fecal coliforms MPN/100 mL 20 × 105 EC 550–900 μS/cm TDS 0.9–1 g/L Color 800–1197 Pt/Co SVI 900–950 mL/g	Fenton process	TS 50% VS 61% COD 53% Color 61% pH = 3 Reaction time 150 min $c(H_2O_2)$ = 4000 mg/L	[40]
TOC 132 mg/L COD 480 mg/L BOD5 267 mg/L Slaughterhouse in Puente Alto, Santiago, Chile	Anaerobic digestion followed by solar photoelectron-Fenton (SPEF)	COD 97% Turbidity and solids total removal CH4 accumulation 90 mL pH 3 Fe^{2+} 1 mM Reaction time 180 min	[41]
pH 6.687 BOD 1078.45 mg/L COD 2024.5 mg/L N 74.8 mg/L Slaughterhouse effluent from Lahore, Pakistan	Photocatalytic oxidation assisted with TiO_2 and silver-doped TiO_2 nanoparticles	BOD 95% COD 87% N 74% Ag-TiO_2-H_2O_2 under UV (400 Watt)	[42]
COD 7.4 g/L TN 609.9 mg/L TP 44.1 mg/L T-Fe 66.9 mg/L Toxic unit (TU) 13.8 Total coliforms 267,000 CFU/ml Slaughterhouse effluent from Nonsan, Korea (average)	Cold plasma	COD 78–93% TN 51–92% TP 35–83% T-Fe 93% Bacteria 98% TU 96%	[43]

Table 5.
Recent application of AOPs on meat processing industry.

amounts of organic materials, suspended solids, oils, salts, and fats [44]. The following AOPs (**Table 6**) are studied to degrade excessive pollutants.

Kinetic models describing the COD degradation of flotation/ozonation processes have been developed and compared [45]. The highest kinetic constant can be achieved for ozonation at acidic medium. The results imply that pre-treatment is necessary to remove the scavengers of HO· in milk. Moreover, participation of casein at low pH might also contribute to the removal of COD. The process could be enhanced further by the addition of H_2O_2 [46], which was then optimized via a CCD (pH, dosage of H_2O_2, ozone, and catalyst). Ozonation is also efficient in decomposing antibiotic in milk based on electrophilic attack by O_3 rather than generated HO· [52].

RSM has been applied to optimize the operation condition of electro-Fenton process with iron electrodes [47]. In order to maximize the removal rate of color and COD, five factors are investigated, including reaction time, current density, pH, H_2O_2/DW (mL/L), and molar ratio. The results reveal strong interactions between pH and molar ratio, as well as between pH and current density. Increasing the current density below the optimal point accelerates Fe^{2+} regenerated from Fe^{3+}, but higher current density results in the generation of O_2 and H_2. Longer reaction times increase removal efficiency, while high pH results in iron ion precipitation, and low pH affects H_2O_2 decomposition into ·OH. Similar process analyzed via RSM was done for Fenton only processes [48]. The electro-Fenton process has the advantage of less consumption of chemicals. Besides removal of chemicals, disinfection of dairy effluent has also been studied [51]. The electrocoagulation process followed by electro-Fenton (EF) or UVA-assisted photoelectro-Fenton (PEF) is suitable for this objective. The results show that EF and PEF are more efficient in inactivation. PEF is better than EF because of UVA radiation.

TiO_2 has also been used as a catalyst to disinfect dairy wastewater [8] which is used as an irrigation water resource. The results show that disinfection efficiency is higher with the addition of TiO_2 and that oxygen accelerates the photocatalysis process. The removal efficiencies of COD, BOD, and SS increased with TiO_2 as photocatalyst illuminated by UV light combined with biological treatment compared with photocatalytic UV reactor only [49].

AOPs in dairy wastewater often operate with biological treatment methods. The operation conditions are optimized by RSM and well explained. Other than elimination of general TOC, COD, and BOD, dairy wastewater treatment also requires disinfection for further application.

3.5 Miscellaneous food industry wastewaters

The diversity of the food industry results in a multitude of wastewater effluents. They are often rich in refractory organics and color compounds. The following table presents existing technologies in other types of food industries (**Table 7**). AOPs have also shown capability for degradation of these effluents.

Molasses used in the production of baker's yeast result in dark color and high organic load of wastewater. The color compound is recalcitrant to aerobic and anaerobic processing. Ultrasonic irradiation assisted by TiO_2-ZnO decolorized the effluent by 25% [53] after optimization (factors: reaction temperature, catalyst composite and calcination temperature, and catalyst load). However, the COD did not change because ultrasonic irradiation transformed organics into smaller intermediate.

In beverage production, 1.72 L wastewater is produced for every 1 L beverage [54]. Photocatalytic processes have been used to degrade synthesized beverage wastewater effluent [54], showing advantages of cerium doped ZnO and favorable

Dairy wastewater origin	Treatment process	Observations	Reference
COD 2000 mg/L Simulate WW: whole milk + distilled water	Flotation/ozonation	More than 80%	[45]
COD 2000 mg/L Simulate WW: whole milk + distilled water	Flotation followed by O_3/H_2O_2 and O_3/Mn^{2+}	64.5% O_3 42.9 mg/L H_2O_2 1071.5 mg/L pH 10.9	[46]
COD 2527 mg/l Color 100 pH 6.27 Conductivity 210 μs/cm TDS 981 mg/L	electro-Fenton process	Factors — COD — Color Removal — 91.76 — 95.22 Current density (mA/cm^2) — 56 — 55.1 Reaction time (min) — 90 — 86 pH — 7.52 — 7.48 Molar ratio H_2O_2/Fe^{2+} — 3.965 — 3.987 H_2O_2/DW (mL/L) — 0.898 — 0.907	[47]
COD 6055 mg/L TS 11900 mg/L TSS 1320 mg/L Color (Pt-Co) 1700 pH 5.7 Conductivity 6 mS/cm Wastewater treatment plant produces milk, yogurt, and butter	Electro-Fenton process	electro-Fenton COD 72% Orthophosphate 88% SS 92% Color removal efficiencies 92.5% H_2O_2/COD ratio 2, current density 32 mA/cm2, pH 2.4 and reaction time 45 min	[48]
pH 6.5–7.5 Turbidity 2–5 NTU DO 6.5–7.5 mg/L Microbes 2300e2900 CFU·mL^{-1} Dairy wastewater after treated by activated sludge process (extended aeration)	Solar photocatalysis (ph-C S) concentrated solar photocatalysis (ph-C CS) solar photolysis (ph-L S) concentrated solar photolysis (ph-L CS) TiO_2 as catalyst	Disinfection efficiency ph-C S 41% ph-C CS 97% ph-L S 10.5% ph-L CS 68.9% Reaction time 30 min	[8]
COD 876 ± 255 mg/L BOD 33 ± 10 mg/L SS 580 ± 159 mg/L Conductivity 1.6 ± 0.4 mS/cm pH 7.9 DO 1.8 ± 0.3 mg/L Dairy effluent after a three-step piggery wastewater treatment (TPWT) system, involving (1) solid/liquid separation, (2) anaerobic treatment, and (3) aerobic	Photocatalytic UV reactor (UVR) followed by bioreactor	Photocatalytic + biological COD 72% BOD 98% SS 79% UV only COD 53% BOD 62% SS 62%	[49]

Dairy wastewater origin	Treatment process	Observations	Reference
treatment (activated sludge basin with a final clarifier)			
Wastewater after anaerobic and aerobic ponds	Ultraviolet/persulfate (UV/PS) oxidation		[50]
pH 5.7 ± 0.2 Conductivity 2.95 ± 0.12 mS/cm TOC 1416 ± 24 mg C L^{-1}	Electrocoagulation (EC) with Fe electrodes followed by electro-Fenton (EF) or UVA-assisted photoelectro-Fenton (PEF) with BDD or RuO_2-based anode	Not available	[51]

Table 6.
Recent application of AOPs in the dairy industry.

Wastewater origin	Treatment process	Observations	Reference
Absorbance at 400 nm 0.3–0.4 pH 5–6 COD 4800–5400 mg/L Baker's yeast factory effluent (Turkey)	Ultrasonic irradiation TiO_2-ZnO as sonocatalyst	Decolorization 25% COD 22.4% (no reduction by ultrasound) Ultrasonic irradiation 20 kHz, 200 W TiO_2/ZnO 4:1 molar ratio (0.15 g/L) 700°C for 60 min	[53]
COD (mg/L) 500–3000 pH 4.7 EC (μs/cm^2) 1845 Turbidity (NTU) 100 Total dissolved solids (TDS) (mg/L) 850 Synthesized beverage industry effluent	UV and solar illumination assisted by synthesized catalyst, immobilized cerium doped ZnO	Photodegradation efficiency* Catalyst dosage 1–5% 35.33–51.56% (UV) 19.86–32.45% (visible light) COD initial concentration 500–3000 mg/L 65.14–21.9% (UV) 42.13–10.12% (visible light) Reaction time 120 min	[54]
COD 15290 ± 855 mg/L TSS 14950 ± 2400 mg/L Palm oil mill effluent (POME) from Rantau, Malaysia	Coagulation by chitosan, addition of ferrous sulfate ($FeSO_4$), chitosan with hydrogen peroxide (H_2O_2), and chitosan with Fenton oxidation	COD 82.82 ± 1.71% TSS 89.92 ± 0.48% Chitosan (2500 mg/L) with H_2O_2 (500 mg/L) pH 7 Reaction time 15 min mixing +1 h sedimentation	[55]
pH 3.6–4.5 Conductivity (uS/cm 20 C) 450–550 TSS (mg/L) 240–280 COD (mg/L) 10,000 BOD_5 (mg/L) 4246–5252 DOC (mg/L) 4218–4260 NTU 130–160 Color 450–100 BOD_5/COD 0.43–0.53 Diluted citrus effluents from Cuba	1. Ozone-based processes (O_3, O_3/OH^-, O_3/UV, O_3/H_2O_2, and $O_3/UV/H_2O_2$) 2. Solar photo-Fenton treatment	1. Ozone-based processes COD 15.7% DOC 10.9% pH 7 Ozone 1.9 g/L UV 254 nm Reaction time 150 min H_2O_2 1017 mg/L 1. Solar photo-Fenton treatment COD 77% DOC 53%	[56]

**Calculation based on adsorption reading through spectrophotometry.*

Table 7.
Recent application of AOPs in miscellaneous food industries.

effect of an acidic environment. Large amounts of wastewater effluents result from the fast growth of the palm oil industry [55]. AOPs utilizing $FeSO_4$ or H_2O_2 and the contribution of chitosan as flocculant have been investigated for such water. Process optimization revealed opposite fluctuation of removal efficiency for COD and TSS with different combination of reagents. Furthermore, chitosan (2500 mg/L) with H_2O_2 (500 mg/L) results in the best removal efficiency as post-treatment of anaerobically digested POME [55].

The effluents of citrus fruit processing facilities are characterized by acidity, presence of essential oils, and toxicity. Diluted real citrus juice wastewater was investigated with various ozonation process, and it was found that the solar-Fenton process is better both in removal efficiency and economic perspective [56].

In the coffee industry, wastewater effluent treatments have recently been reviewed elsewhere [57]. The main characteristics of these streams are the presence of various color compounds and macromolecules. AOPs have been proven to be efficient in combination with biological treatment [58]. However, considering the nature of color compounds, ion exchange is a more promising field.

Wastewater treatment for different food or beverages should be adjusted according to the characteristics of different effluents such as complex color compounds (yeast and coffee) or low pH (juice). Pre-treatment such as coagulation or anaerobic digestion is favorable.

3.6 Food dyes

Apart from wastewater effluents from food processing factories directly, municipal wastewater contains some refractory compounds from food industries. Food dyes are extensively used in fruit juices and sweets products as food additives. The colored effluents from such industries need to be treated carefully before discharge. Otherwise they will lead to issues such as pollution on esthetic grounds and interference of light transmission [59].

Food dye	Treatment process	K_{app}	Reference
Ponceau 4R	Electro-oxidation 1. Electrogenerated H_2O_2 ($EO–H_2O_2$) 2. Electro-Fenton (EF) 3. Photoelectro-Fenton (PEF)	(10^{-2} min^{-1}) 1. 2.72 ± 0.41 2. 12.31 ± 0.53 3. 13.35 ± 0.66 (real water matrices)	[61]
Amaranth food dye (AM)	Heterogeneous E-Fenton process with synthesized $Fe_3 - xCu_xO_4$ ($0 \leq x \leq 0.25$) NPs	4.2×10^{-2} min^{-1} (x = 0.25)	[62]
Brilliant blue FCF (BBF)	Fe_3O_4-TiO_2(FTNs) assisted with different UV light	0.059 min^{-1} (FTNs/UVA/PMS)	[63]
Carmoisine (E22)	UVA-LEDs/PMS/Fe^{2+}	0.1553 min^{-1}	[64]
Tartrazine	Visible light photo-Fenton oxidation with three bismuth oxyhalide catalysts	0.0026–0.06 min^{-1} Temperature from 30–70°C	[65]
Sunset yellow FCF (SY)	Electrochemical assisted with palladium-ruthenium nanoparticles incorporated with carbon aerogel (Pd-Ru/CA)	0.295 s^{-1}	[66]
Tartrazine yellow (TT) and brilliant blue (BB)	Photocatalytic process	0.016 min^{-1}	[67]

Table 8.
Recent application of AOPs in food dyes.

Azo compounds are widely used in food industries because of its brilliant shades, relative low cost, and simple manufacture [60]. However, azo compounds are resistant to conventional treatment due to one or more azo bonds [61]. Hence the following AOPs (**Table 8**) are applied to degrade specific food dyes.

Fenton process is widely used. Thiam et al. proposed the routes for ponceau 4R degradation. The Fenton process was found to provide fast ·OH production and stability despite the interference of real water matrices [61].

Fenton processes can be enhanced by various catalysts. In order to circumvent the pH limitation of Fenton processes, catalysts with immobilized iron ions are synthesized to avoid precipitation. The combination of magnetite (Fe_3O_4) and copper presents high specific surface area and synergic effects between Cu^{2+}/Cu^{+} and Fe^{3+}/Fe^{2+}. Moreover, it is easy to separate [62]. Similarly, Fe_3O_4-TiO_2 (FTNs) [63] as well as metal doped bismuth oxyhalide catalysts (BiOCl, Cu-BiOCl, and Fe-BiOCl) have been used to degrade BBF [65].

Compared with conventional UVA, UVA-LED are more efficient and cost beneficial. The implementation of UVA-LED accelerates the regeneration of radicals. [64]. Zazouli et al. also suggest that UVC, a high energy UV source, is more efficient in decomposing PMS into ·OH and $SO_4{\cdot}^-$ than UVA [63]. The replacement of batch reactors by flow reactors is another promising modification [67].

In dealing with a specific food dye, catalysts are synthesized to assist AOPs. The operation conditions such as catalyst dosage and UV source have to be considered. Further studies are required related to economic catalyst development and ease of separation and recovery.

4. Practical application of AOPs

Wastewater treatment in the food industry with AOPs is a flourishing field. Nevertheless, pilot-scale plants are not widely studied. A number of studies implement AOPs in sequential approaches with a complete solution for the treatment and reuse of a complex wastewater from food industries.

Yalılı Kılıç et al. investigated the pilot-scale treatment of olive oil mill wastewater [68]. The combination of physicochemical treatment, ultrafiltration, and $O_3/H_2O_2/UV$ presented the most favorable results of pollutant removal. However, cost evaluation indicates that the absence of ozone is more economic without removal efficiency diminishment. Another study applied Fenton process in a CSTR at pilot scale [69]. The results highlight the importance of operation condition optimization in actual plant condition. However, such a process is not affordable for small facilities generating large amount of pollutants.

In the winery wastewater treatment field, ozone-based AOPs (O_3/UV and $O_3/UV/H_2O_2$) at pilot-scale bubble column reactors have shown high efficiency in TOC removal. Moreover, addition of H_2O_2 is favorable considering overall costs [70]. One of the origins of winery wastewater is cork boiling water. AOPs including solar photo-Fenton and ozone have been applied to the effluents pre-treated by physicochemical methods. [71] The pre-treatment reduces the additional benefits of AOPs such as increased biodegradability and toxicity reduction. Further assessments of overall costs remain to be investigated.

5. Conclusion

In the food industry, wastewater effluents including complex organics are generated from various food processing steps and equipment maintenance procedures.

Moreover, the concentration of containments varies according to the type of food. This situation leads to an urgent demand for AOPs as complementary technologies to traditional wastewater treatment which is insufficient to process excess pollutants. The effluents from AOPs are more biodegradable for biological treatments, and hence the addition of AOPs as pre-treatment or post-treatment is a promising and economical solution for processing various wastewaters in the food industry.

Further research on catalysts that increase the amounts of effective radicals, simple separation procedure, and high recovery rates is suggested. Reactor design and optimization is another promising field. Due to the small number of practical applications, additional pilot-scale studies with AOPs are also recommended. Large-scale studies can provide overall cost evaluation including capital cost, operation cost, and possible profits from by-products. Such integrated economic assessment in real plants will be valuable guidance for future research.

Author details

Zhaoran Xin and Lars Rehmann*
Department of Chemical and Biochemical Engineering, Western University London, ON, Canada

*Address all correspondence to: lrehmann@uwo.ca

IntechOpen

© 2020 The Author(s). Licensee IntechOpen. Distributed under the terms of the Creative Commons Attribution - NonCommercial 4.0 License (https://creativecommons.org/licenses/by-nc/4.0/), which permits use, distribution and reproduction for non-commercial purposes, provided the original is properly cited. (cc) BY-NC

References

[1] Krzemińska D, Neczaj E, Borowski G. Advanced oxidation processes for food industrial wastewater decontamination. Journal of Ecological Engineering. 2015;**16**:61-71

[2] Ghosh Ray S, Ghangrekar MM. Comprehensive review on treatment of high-strength distillery wastewater in advanced physico-chemical and biological degradation pathways. International Journal of Environmental Science and Technology. 2019;**16**: 527-546

[3] Lima VN, Rodrigues CSD, Borges RAC, et al. Gaseous and liquid effluents treatment in bubble column reactors by advanced oxidation processes: A review. Critical Reviews in Environmental Science and Technology. 2018;**48**: 949-996

[4] Water and Wastewater Monitoring Report. Region of Waterloo [online]. Available from: https://www.regionofwaterloo.ca/en/living-here/resources/Documents/water/reports/WS2019WaterAndWastewater Monitoring Report.pdf

[5] The UV/Oxidation Handbook, Solarchem Environmental Systems. Markham, Ontario; 1994

[6] Staehelin J, Holgné J. Decomposition of ozone in water: Rate of initiation by hydroxide ions and hydrogen peroxide. Environmental Science & Technology. 1982;**16**:676-681

[7] Carey JH. An introduction to advanced oxidation processes (AOP) for destruction of organics in wastewater. Water Quality Research Journal. 1992;**27** (1):1-22

[8] Afsharnia M, Kianmehr M, Biglari H, et al. Disinfection of dairy wastewater effluent through solar photocatalysis processes. Water Science and Engineering. 2018;**11**:214-219

[9] Heponiemi A, Lassi U. Advanced oxidation processes in food industry wastewater treatment – A review. In: Food Industrial Processes - Methods and Equipment. London, UK: InTech; 2012

[10] Azbar N, Bayram A, Filibeli A, et al. A review of waste management options in olive oil production. Critical Reviews in Environmental Science and Technology. 2004;**34**:209-247

[11] Union PO of the E. Survey on current activity on the valorization of by-products from the olive oil industry

[12] Galliou F, Markakis N, Fountoulakis MS, et al. Production of organic fertilizer from olive mill wastewater by combining solar greenhouse drying and composting. Waste Management. 2018; **75**:305-311

[13] Elkacmi R, Bennajah M. Advanced oxidation technologies for the treatment and detoxification of olive mill wastewater: A general review. Journal of Water Reuse and Desalination. 2019;**9**: 463-505

[14] Lofrano G, Meric S. A comprehensive approach to winery wastewater treatment: A review of the state-of the-art. Desalination and Water Treatment. 2016;**57**:3011-3028

[15] Ioannou LA, Puma GL, Fatta-Kassinos D. Treatment of winery wastewater by physicochemical, biological and advanced processes: A review. Journal of Hazardous Materials. 2015;**286**:343-368

[16] Pulido JMO. A review on the use of membrane technology and fouling control for olive mill wastewater treatment. Science of the Total Environment. 2016;**563–564**:664-675

[17] Ioannou-Ttofa L, Michael-Kordatou I, Fattas SC, et al. Treatment efficiency

and economic feasibility of biological oxidation, membrane filtration and separation processes, and advanced oxidation for the purification and valorization of olive mill wastewater. Water Research. 2017;**114**:1-13

[18] European IPPC Bureau. Best Available Techniques Reference document on food, drink and milk industries. 2019. Available from: http://eippcb.jrc.ec.europa.eu/reference/brefdownload/download_FDM.cfm

[19] WHO. Total alcohol per capita (15+ years) consumption, in litres of pure alcohol. WHO

[20] Petruccioli M, Duarte JC, Federici F. High-rate aerobic treatment of winery wastewater using bioreactors with free and immobilized activated sludge. Journal of Bioscience and Bioengineering. 2000;**90**:381-386

[21] Díez AM, Iglesias O, Rosales E, et al. Optimization of two-chamber photo electro Fenton reactor for the treatment of winery wastewater. Process Safety and Environment Protection. 2016;**101**: 72-79

[22] Díez AM, Rosales E, Sanromán MA, et al. Assessment of LED-assisted electro-Fenton reactor for the treatment of winery wastewater. Chemical Engineering Journal. 2017;**310**:399-406

[23] Candia-Onfray C, Espinoza N, Sabino da Silva EB, et al. Treatment of winery wastewater by anodic oxidation using BDD electrode. Chemosphere. 2018;**206**:709-717

[24] Rodríguez-Chueca J, Amor C, Mota J, et al. Oxidation of winery wastewater by sulphate radicals: Catalytic and solar photocatalytic activations. Environmental Science and Pollution Research. 2017;**24**:22414-22426

[25] Amor C, Rodríguez-Chueca J, Fernandes JL, et al. Winery wastewater treatment by sulphate radical based-advanced oxidation processes (SR-AOP): Thermally vs UV-assisted persulphate activation. Process Safety and Environment Protection. 2019;**122**: 94-101

[26] Guimarães V, Lucas MS, Peres JA. Combination of adsorption and heterogeneous photo-Fenton processes for the treatment of winery wastewater. Environmental Science and Pollution Research. 2019;**26**:31000-31013

[27] Malandra L, Wolfaardt G, Zietsman A, et al. Microbiology of a biological contactor for winery wastewater treatment. Water Research. 2003;**37**: 4125-4134

[28] Artiga P, Ficara E, Malpei F, et al. Treatment of two industrial wastewaters in a submerged membrane bioreactor. Desalination. 2005;**179**:161-169

[29] Bustamante MA, Moral R, Paredes C, et al. Agrochemical characterisation of the solid by-products and residues from the winery and distillery industry. Waste Management. 2008;**28**:372-380

[30] Amor C, Marchão L, Lucas MS, et al. Application of advanced oxidation processes for the treatment of recalcitrant agro-industrial wastewater: A review. Water. 2019;**11**(2):205

[31] Mekki A, Dhouib A, Sayadi S. Review: Effects of olive mill wastewater application on soil properties and plants growth. International Journal of Recycling of Organic Waste in Agriculture. 2013;**2**:15

[32] García CA, Hodaifa G. Real olive oil mill wastewater treatment by photo-Fenton system using artificial ultraviolet light lamps. Journal of Cleaner Production. 2017;**162**:743-753

[33] Amaral-Silva N, Martins RC, Nunes P, et al. From a lab test to industrial

application: Scale-up of Fenton process for real olive mill wastewater treatment. Journal of Chemical Technology and Biotechnology. 2017;**92**:1336-1344

[34] Ates H, Dizge N, Cengiz YH. Combined process of electrocoagulation and photocatalytic degradation for the treatment of olive washing wastewater. Water Science and Technology. 2017;**75** (1):141-154

[35] Kirmaci A, Duyar A, Akgul V, et al. Optimization of combined ozone/ Fenton process on olive mill wastewater treatment. Aksaray University Journal of Science and Engineering. 2018;**2**: 52-62

[36] Genç N, Durna E, Kayapinar Cicigün HK. Response surface modeling and optimization of microwave-activated Persulfate oxidation of olive oil mill wastewater. Clean: Soil, Air, Water. 2020;**48**:1-11

[37] Bustillo-Lecompte C, Mehrvar M, Quiñones-Bolaños E. Slaughterhouse wastewater characterization and treatment: An economic and public health necessity of the meat processing industry in Ontario, Canada. International Conference on Environmental Pollution and Public Health. 2016;**2016**:175-186

[38] Bustillo-Lecompte CF, Mehrvar M. Treatment of an actual slaughterhouse wastewater by integration of biological and advanced oxidation processes: Modeling, optimization, and cost-effectiveness analysis. Journal of Environmental Management. 2016;**182**: 651-666

[39] Naderi KV, Bustillo-Lecompte CF, Mehrvar M, et al. Combined UV-C/ H2O2-VUV processes for the treatment of an actual slaughterhouse wastewater. Journal of Environmental Science and Health Part B Pesticides Food Contaminants and Agricultural Wastes. 2017;**52**:314-325

[40] Masoumi Z, Shokohi R, Atashzaban Z, et al. Stabilization of excess sludge from poultry slaughterhouse wastewater treatment plant by the Fenton process. Avicenna Journal of Environmental Health Engineering. 2015;**2**:3239-3239

[41] Vidal J, Huiliñir C, Salazar R. Removal of organic matter contained in slaughterhouse wastewater using a combination of anaerobic digestion and solar photoelectro-Fenton processes. Electrochimica Acta. 2016;**210**:163-170

[42] Bukhari K, Ahmad N, Sheikh IA, et al. Effects of different parameters on Photocatalytic oxidation of slaughterhouse wastewater using TiO2 and silver-doped TiO2Nanoparticles. Polish Journal of Environmental Studies. 2019;**28**:1591-1600

[43] Kim HJ, Won CH, Kim HW. Pathogen deactivation of glow discharge cold plasma while treating organic and inorganic pollutants of slaughterhouse wastewater. Water, Air, and Soil Pollution. 2018;**229**:1-10

[44] Sánchez IMR, Ruiz JMM, López JLC, et al. Effect of environmental regulation on the profitability of sustainable water use in the agro-food industry. Desalination. 2011;**279**:252-257

[45] Carvalho MCS, Borges AC, Pereira MDS, et al. Degradation kinetics of organic matter in dairy industry wastewater by flotation/ozonation processes. Bioscience Journal. 2018;**34**: 587-594

[46] Maroneze MM, Zepka LQ, Vieira JG, et al. A tecnologia de remoção de fósforo: Gerenciamento do elemento em resíduos industriais. Revista Ambiente & Agua. 2018;**9**:445-458

[47] Davarnejad R, Nikseresht M. Dairy wastewater treatment using an electrochemical method: Experimental and statistical study. Journal of

Electroanalytical Chemistry. 2016;**775**: 364-373

[48] Akkaya GK, Erkan HS, Sekman E, et al. Modeling and optimizing Fenton and electro-Fenton processes for dairy wastewater treatment using response surface methodology. International Journal of Environmental Science and Technology. 2019;**16**:2343-2358

[49] Su JC, Wang YL, Su JJ. Photocatalytic oxidation of dairy effluent with UV lamp or UV light-emitting diode module and biological treatment processes. International Journal of Environmental Science and Technology. 2019;**16**:1047-1056

[50] Pramanik BK, Hai FI, Roddick FA. Ultraviolet/persulfate pre-treatment for organic fouling mitigation of forward osmosis membrane: Possible application in nutrient mining from dairy wastewater. Separation and Purification Technology. 2019;**217**:215-220

[51] Bruguera-Casamada C, Araujo RM, Brillas E, et al. Advantages of electro-Fenton over electrocoagulation for disinfection of dairy wastewater. Chemical Engineering Journal. 2019; **376**:119975

[52] Alsager OA, Alnajrani MN, Abuelizz HA, et al. Removal of antibiotics from water and waste milk by ozonation: Kinetics, byproducts, and antimicrobial activity. Ecotoxicology and Environmental Safety. 2018;**158**:114-122

[53] Yılmaz E, Fındık S. Sonocatalytic treatment of baker's yeast effluent. Journal of Water Reuse and Desalination. 2017;7:88-96

[54] Vakili B, Shahmoradi B, Maleki A, et al. Synthesis of immobilized cerium doped ZnO nanoparticles through the mild hydrothermal approach and their application in the photodegradation of synthetic wastewater. Journal of Molecular Liquids. 2019;**280**:230-237

[55] Parthasarathy S, Gomes RL, Manickam S. Process intensification of anaerobically digested palm oil mill effluent (AAD-POME) treatment using combined chitosan coagulation, hydrogen peroxide (H_2O_2) and Fenton's oxidation. Clean Technologies and Environmental Policy. 2016;**18**:219-230

[56] Guzmán J, Mosteo R, Sarasa J, et al. Evaluation of solar photo-Fenton and ozone based processes as citrus wastewater pre-treatments. Separation and Purification Technology. 2016;**164**: 155-162

[57] Ijanu EM, Kamaruddin MA, Norashiddin FA. Coffee processing wastewater treatment: A critical review on current treatment technologies with a proposed alternative. Applied Water Science. 2020;**10**:1-11

[58] Ibarra-Taquez HN, GilPavas E, Blatchley ER, et al. Integrated electrocoagulation-electrooxidation process for the treatment of soluble coffee effluent: Optimization of COD degradation and operation time analysis. Journal of Environmental Management. 2017;**200**:530-538

[59] Shaikh IA, Munir S, Ahmed F, et al. Advanced oxidative removal of C.l. food red 17 dye from an aqueous solution. Pakistan Journal of Nutrition. 2014;**13**: 631-634

[60] Thiam A, Brillas E. Treatment of a mixture of food color additives (E122, E124 and E129) in different water matrices by UVA and solar photoelectro-Fenton. Water Research. 2015;**81**:178-187

[61] Thiam A, Brillas E, Garrido JA, et al. Routes for the electrochemical degradation of the artificial food azo-colour Ponceau 4R by advanced oxidation processes. Applied Catalysis B: Environmental. 2016;**180**:227-236

[62] Barros WRP, Steter JR, Lanza MRV, et al. Catalytic activity of Fe3-xCuxO4

(0≤x≤0.25) nanoparticles for the degradation of Amaranth food dye by heterogeneous electro-Fenton process. Applied Catalysis B: Environmental. 2016;**180**:434-441

[63] Zazouli MA, Ghanbari F, Yousefi M, et al. Photocatalytic degradation of food dye by Fe3O4-TiO2 nanoparticles in presence of peroxymonosulfate: The effect of UV sources. Journal of Environmental Chemical Engineering. 2017;**5**:2459-2468

[64] Ahmadi M, Ghanbari F, Alvarez A, et al. UV-LEDs assisted peroxymonosulfate/Fe2+ for oxidative removal of carmoisine: The effect of chloride ion. Korean Journal of Chemical Engineering. 2017;**34**:2154-2161

[65] Tekin G, Ersöz G, Atalay S. Visible light assisted Fenton oxidation of tartrazine using metal doped bismuth oxyhalides as novel photocatalysts. Journal of Environmental Management. 2018;**228**:441-450

[66] Thirumalraj B, Rajkumar C, Chen SM, et al. Carbon aerogel supported palladium-ruthenium nanoparticles for electrochemical sensing and catalytic reduction of food dye. Sensors and Actuators B: Chemical. 2018;**257**:48-59

[67] Júnior WJ, Júnior N, Aquino RVS, et al. Development of a new PET flow reactor applied to food dyes removal with advanced oxidative processes. Journal of Water Process Engineering. 2019;**31**:100823

[68] Yalılı Kılıç M, Yonar T, Kestioğlu K. Pilot-scale treatment of olive oil mill wastewater by physicochemical and advanced oxidation processes. Environmental Technology. 2013;**34**: 1521-1531

[69] Hodaifa G, Ochando-Pulido JM, Rodriguez-Vives S, et al. Optimization of continuous reactor at pilot scale for olive-oil mill wastewater treatment by Fenton-like process. Chemical Engineering Journal. 2013;**220**:117-124

[70] Lucas MS, Peres JA, Li PG. Treatment of winery wastewater by ozone-based advanced oxidation processes (O3, O3/UV and O3/UV/ H2O2) in a pilot-scale bubble column reactor and process economics. Separation and Purification Technology. 2010;**72**:235-241

[71] De Torres-Socías E, Fernández-Calderero I, Oller I, et al. Cork boiling wastewater treatment at pilot plant scale: Comparison of solar photo-Fenton and ozone (O3, O3/H2O2). Toxicity and biodegradability assessment. Chemical Engineering Journal. 2013;**234**:232-239

Chapter 7

Catalytic Ozone Oxidation of Petrochemical Secondary Effluent: Mechanism, Application and Future Development

Yu Tan, Liya Fu, Changyong Wu, Yanan Li, Xiumei Sun and Yuexi Zhou

Abstract

Petrochemical secondary effluent contains toxic and refractory organic compounds, which are difficult to be further treated by traditional biological process. In China, most of the advanced treatment units have been built recently by catalytic ozone oxidation process to achieve the high-quality effluent. In this chapter, the mechanism and reaction process of catalytic ozone oxidation of petrochemical secondary effluent will be introduced in detail. With the operation of the catalytic ozone oxidation tank, a series of problems which are not taken into account at the beginning of the design have arisen. The chapter will talk about the problems concerning the biological flocs, colloidal macromolecule organic compounds, ozone mass transfer, and catalysts based on practical applications. In the last part of the chapter, the development trends of catalytic ozone oxidation of petrochemical secondary effluent will also be discussed.

Keywords: ozone, petrochemical wastewater, reaction mechanism, biological flocs, trends

1. Introduction

As an important part of the chemical industry, the petrochemical industry is one of the pillar industries in China and has a large proportion in the national economy. According to China Industrial Research Network, the total output value of China's petrochemical industry in 2014 has exceeded 13 trillion CNY, ranking second in the world. In 2016, the total discharge of petrochemical wastewater reached 2 billion tons, accounting for more than 10% of the national industrial wastewater discharge. The organic pollution wastewater generated during the operation of petrochemical enterprises is an important part of China's industrial wastewater. Types of pollutants, large concentration, high toxicity, as well as containing many refractory biodegradation organic substances make water quality of the petrochemical wastewater complex; its governance has become a bottleneck restricting the development of the petrochemical industry.

Complex water quality features have spawned new water treatment processes. Advanced oxidation processes (AOPs) can produce hydroxyl radicals (•OH) with strong oxidizing ability under the reaction conditions of electricity, sound, photo, catalyst, and others. Hydroxyl radical (•OH) is an important reactive oxygen species with a standard oxidation–reduction potential of +2.8 V, which is second only to fluorine in nature [1]. It can successfully decompose most pollutants including refractory organics in water and oxidize macromolecule refractory organic matter into low-toxic or nontoxic small molecular substances.

The catalytic ozone oxidation process is a kind of advanced oxidation process, which has the following characteristics: (a) it produces strong oxidizing •OH, and the oxidation process is fast, reaching 10^6 to 10^{10}L/ (mol•s); (b) it performs excellently in removing refractory organic matter with no secondary pollutants, which means the process is green and efficient; and (c) the process runs economically. Taking a petrochemical integrated sewage treatment plant in northern China as an example, ozone can directly use air as raw material by ozone generator without adding auxiliary materials, and the earliest used catalyst has been running normally for 5 years, which still has high catalytic activity now. Therefore, catalytic ozone oxidation process is widely used in the advanced treatment of petrochemical wastewater. Especially after July 1, 2015, the Ministry of Ecology and Environment began to implement the *Petrochemical industry pollutant discharge standards (GB 31571-2015)* to replace the *Integrated Wastewater Discharge Standard (GB 8978-1996),* which improves the direct drainage's COD limit of sewage treatment plant in petrochemical industrial park from 100 to 60 mg/L and adds 60 indicators of characteristic organic pollutants. Catalytic ozone oxidation process is increasingly popular in wastewater treatment due to the advantage of high efficiency, low consumption, energy saving, and environmental protection.

2. Research progress of catalytic ozone oxidation

Due to the excellent performance of catalytic ozone oxidation process in industrial wastewater treatment, substantial researches on the development of new catalysts and catalytic processes have been carried out. But so far the mechanism of catalytic ozone oxidation has not been unified in the academia, even the mechanisms proposed in many published literature are contradictory, which has largely limited the promotion of catalytic ozone oxidation process as an efficient, green technology. Several possible mechanisms on the catalytic ozone oxidation process will be introduced in the following.

According to the state of the catalyst, the catalytic ozone oxidation process can be divided into homogeneous and heterogeneous catalytic oxidation.

2.1 Homogeneous catalytic oxidation

Some transition metal cations have been shown to improve ozonation efficiency; among the most widely used are Mn(II), Fe(III), Fe(II), Co(II), Cu(II), Zn(II), and Cr(III) [2–4]. In simple terms, homogeneous catalytic oxidation is to let the transition metal ion decompose ozone or assist organics to be more easily oxidation. Two major mechanisms of homogeneous catalytic ozone oxidation can be found in the published research papers:

1. Ozone decomposes under the action of the catalyst to generate free radicals [5–7].

2. A complex coordination reaction occurs between the catalyst and the organic matter or ozone to promote the reaction between ozone and organic matter [2, 8, 9].

The free radical mechanism is similar to the Fenton reaction, and the electron transfer is achieved by oxidation and reduction of the valence metal ions. Based on the early work of Logager et al. [10], Sauleda et al. [7] explained the formation of hydroxyl radicals in the presence of Fe(II) ions:

$$Fe^{2+} + O_3 \rightarrow FeO^{2+} + O_2 \quad (1)$$

$$FeO^{2+} + H_2O \rightarrow Fe^{3+} + \bullet OH + OH^- \quad (2)$$

Another mechanism is proposed by Pines and Reckhow [8] on the research catalytic ozone oxidation of oxalic acid in the presence of Co(II) ions. In general, hydroxyl radicals are more likely to be produced under alkaline conditions. However, experiments have shown that the reaction became inefficient with the increase of pH. Moreover in the presence of t-butanol (a radical scavenger), the reaction rate was also found not to change. These suggest that the free radical mechanism is not responsible for mineralization of oxalate in the Co(II)/O_3 system. So the hypothesis illustrated in **Figure 1** is proposed to explain the experimental phenomenon. Mineralization of oxalate is achieved by forming complexes with Co(II) ions.

Other metal ions have also been found to have similar catalytic effects. However, once metal ions are used as catalysts to promote ozone oxidation, the question on how to separate catalysts from the treated water should take into consideration. Therefore, there have been few reports on its research recently.

2.2 Heterogeneous catalytic oxidation

Among the most widely used catalysts in heterogeneous catalytic ozone oxidation are [11]:

- Metal oxides (MnO_2, TiO_2, Al_2O_3, and FeOOH).
- Metals (Cu, Ru, Pt, Co) on supports (SiO_2, Al_2O_3, TiO_2, CeO_2, and activated carbon).
- Activated carbon.

Adsorption is the key word in heterogeneous catalytic ozone oxidation. There are three possible ways which could be the general mechanism of heterogeneous catalytic ozone oxidation (**Figure 2**) [12]:

- Ozone is adsorbed on the surface of the catalyst.
- Organic molecule is adsorbed on the surface of the catalyst.
- Both, ozone and organic molecule, are adsorbed on the catalyst surface.

A lot of research work has been carried out for different catalysts.

2.2.1 Metal oxides

A variety of metal oxides have been shown to be effective in heterogeneous catalytic oxidation, such as MnO_2, Al_2O_3, TiO_2, and FeOOH. All published articles

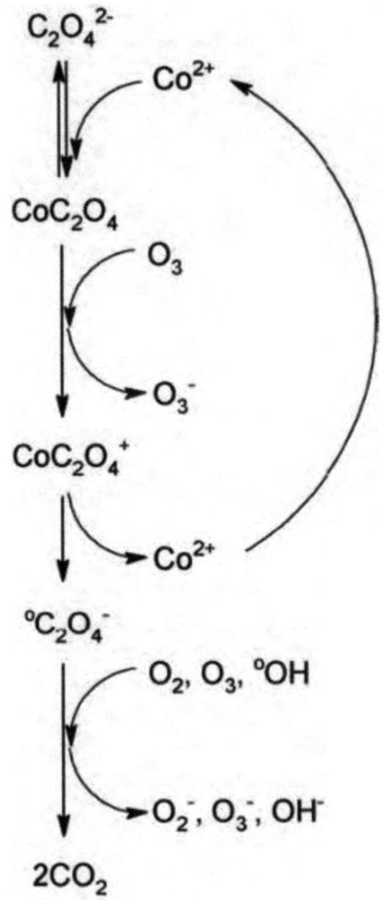

Figure 1.
Mechanism of oxalic acid oxidation in the presence of Co(II) ions proposed by Pines and Reckhow (modified from [8]).

have indicated that the mechanism of catalysis has a tight affinity with the surface chemistry of the oxides. More specifically, metal oxides as catalysts are surrounded by water in wastewater treatment. There is no doubt that the surface is covered by hydroxyls which have ion-exchange properties. Some of the oxides may also have Lewis acid centers which are considered as another adsorption site on their surfaces. In addition, hydrophobic sites can also be spotted on the surface on some metal oxides to explain the adsorption of organic molecules. **Table 1** lists the metal oxides commonly used as catalysts in the research and target organics for removal.

2.2.1.1 MnO_2

In the research of the catalytic mechanism of MnO_2, Andreozzi et al. [33] first found in the study that the catalytic activity of MnO_2 to degrade oxalic acid was the highest when pH = 3.2. This result, however, is contradictory with what we have known: the optimum pH for dissociating ozone to produce •OH is 5–6. This means during the process •OH is not the dominated factor for removing oxalic acid. According to Tong et al. [58], adsorption of organic molecule on the surface of MnO_2 and subsequent attack of ozone on adsorbed organic molecule are responsible for catalytic activity.

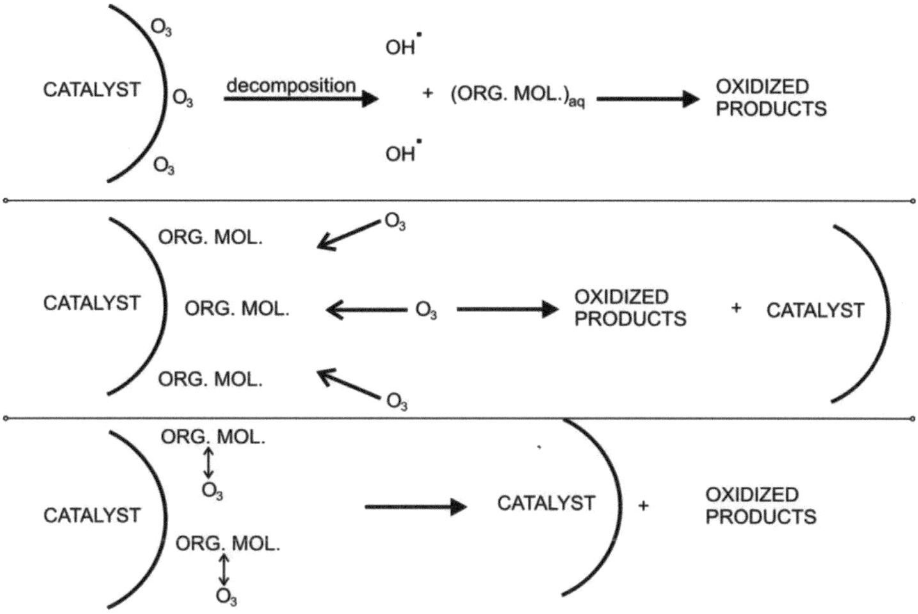

Figure 2.
Three possible cases of heterogeneous catalysis [12].

On the other hand, when atrazine was used as the characteristic pollutant, adding in t-butanol could inhibit the catalytic activity of MnO_2 [14]. If atrazine is replaced with nitrobenzene, t-butanol has light effect on catalytic activity of MnO_2 [59]. From the research, we can draw a conclusion that the catalytic mechanism of MnO_2 catalyzed ozone oxidation to degrade organic matter in water varies depending on the type of organic matter, including the mechanism of free radicals produced by decomposing ozone, the surface coordination mechanism of organic matters, and the combination of both.

2.2.1.2 Al_2O_3

Al_2O_3 is another metal oxide that is widely used as a catalyst. There is a lot of controversy about research of the catalytic mechanism of Al_2O_3. The researches of Kasprzyk-Hordern and Nawrocki had found that in the presence of Al_2O_3, there was no catalytic activity for the removal of aromatic hydrocarbons and ethers. However, in the latter experiments, it was detected that the decomposition rate of ozone is twice that of ozonation alone [60–62]. In Kasprzyk-Hordern [19] another experiment, high catalytic activity was demonstrated when the substrate is natural organic matter (NOM) in the presence of alumina. The key factor to explain this contrast is the capacity of the surface to adsorb different organic molecules, as both hydrocarbons and ethers do not adsorb on alumina from aqueous solutions while alumina revealed high adsorption capacity toward NOM. These could be the evidence of the hypothesis that the catalytic activity of Al_2O_3 depends on the surface absorption of the substrate. However, Ernst et al. [20] reported that the adsorption of organic molecules is detrimental to the catalytic activity of γ-Al_2O_3. The research showed it was the high adsorption of oxalic acid on the surface that caused the lower efficiency. Obviously, the mechanism is remained to be discussed further and deeper.

Catalyst	Target organics	References
MnO_2	Carboxylic acids (oxalic, pyruvic, sulfosalicylic, propionic, glyoxylic), N-methyl-p-aminophenol, atrazine	[3, 8, 13–17]
Al_2O_3	Carboxylic acids (oxalic, acetic, salicylic, succinic), 2-chlorophenol, chloroethanol, NOM, dimethyl phthalate	[18–23]
TiO_2	Oxalic acid, carbamazepine, naproxen, nitrobenzene, clofibric acid	[24–27]
FeOOH	p-Chlorobenzoic acid, NOM	[28–32]
TiO_2, Al_2O_3, Ni_2O_3, CuO, MoO_3, CoO, Fe_2O_3	m-Dinitrobenzene	[33]
ZnO	p-Chlorobenzoic acid	[34, 35]
$TiO_2(2.5\%)/\alpha\text{-}Al_2O_3$	NOM	[36–38]
$TiO_2(10\%)/Al_2O_3$, $Fe_2O_3(10\%)/Al_2O_3$	Oxalic acid, chloroethanol, chlorophenol	[19]
TiO_2/Al_2O_3	Fulvic acids	[38]
$TiO_2(15\%)/\gamma\text{-}Al_2O_3$	Oxalic acid	[39]
$MnO_2(10\%)/TiO_2$	Phenol	[2]
$MnO_x(10.8\%)/AC$	Nitrobenzene	[40]
$\beta\text{-}Al_2O_3$	Pyruvic acid	[41]
$\gamma\text{-}Al_2O_3$	Methylisoborneol	[42]
γ-AlOOH, $\alpha\text{-}Al_2O_3$	2,4,6-trichloroanisole	[43]
MnO_2	Phenol, NMO	[44, 45]
CeO_2	Aniline, sulfanilic acid, dyes	[46]
Co_3O_4/Al_2O_3	Pyruvic acid	[41]
CuO/Al_2O_3	Alachlor, oxalic acid, substituted phenols	[3, 47, 48]
$Co(OH)_2$	p-Chloronitrobenzene	[49]
TiO_2/ AC	Methylene blue	[50]
NiO/Al_2O_3	Oxalic acid	[51]
NiO/CuO	Dichloroacetic acid	[52]
MnO_x/Al_2O_3	Phenazone, ibuprofen, phenytoin, diphenhydramine	[53]
MgO	Dye	[54]
CoO_x/ZrO_2	2,4-D	[55]
Co/CeO, Ag/CeO, Mn/CeO	Phenolic wastewater	[56, 57]

Table 1.
Common catalyst components and target organics for removal (modified from [11]).

2.2.1.3 TiO_2

TiO_2 is usually used as a photocatalyst [63–68], which also has a good catalytic effect when degrading organic matter in wastewater. The catalytic mechanism of TiO_2 has close affinity with the construction of TiO_2.

TiO_2, which is often used as a catalyst, is classified into rutile type and anatase type. Yang et al. [69] used rutile TiO_2 for catalytic ozone oxidation research and found that the efficiency of removing organic matter decreased after adding TBA to the reaction system. This implies that the free radical mechanism dominates the reaction process. On the contrary, Ye et al. [70] studied the degradation of

4-chloronitrobenzene (CNB) during catalytic and photocatalytic ozone oxidation in the presence of TiO_2. The result of the experiment is that the TOC is higher under UV condition, which does not correspond with the free radical mechanism. So far, the difference in this mechanism is still lacking in-depth explanation.

2.2.1.4 FeOOH

The mechanism of FeOOH is similar to the mechanism of homogeneous catalytic oxidation. The difference between them is that the decomposition and oxidation happen in bulk water in homogeneous catalytic oxidation, while for FeOOH, the process occurs on the surface.

1. The process satisfies the hydroxyl radical mechanism. As is illustrated in **Figure 3**, the ozone molecule first forms a surface ring with the hydroxyl group on the surface of the catalyst and then decomposes to produce Fe-OH-O^- and O_2. Finally, Fe-OH-O^- is directly attacked by ozone to form •OH and $O_2•^-$, and the water molecules complex on the surface of FeOOH to form a new hydroxyl group.

2. Organic compounds form complexes on the surface of FeOOH and are then directly oxidized by ozone molecules to degrade organic matter [72].

2.2.2 Metals on supports

Metals on supports are a type of catalysts which use supports (SiO_2, Al_2O_3, TiO_2, CeO_2, and activated carbon) to load metals (Cu, Ru, Pt, Co). They have been observed to have a high catalytic activity in catalytic ozone process.

Two main routes have been proposed to explain the catalytic mechanism [73] (**Figure 4**):

1. The metal supported on the catalyst has a constant valence state in the catalytic reaction and only functions as a coordination.

2. The metals are attacked by the ozone to generate hydroxyl radicals. Then, through the adsorption and desorption of organic matter on the surface of the catalyst, electron transfer is completed to achieve metal reduction and oxidation of organic matter.

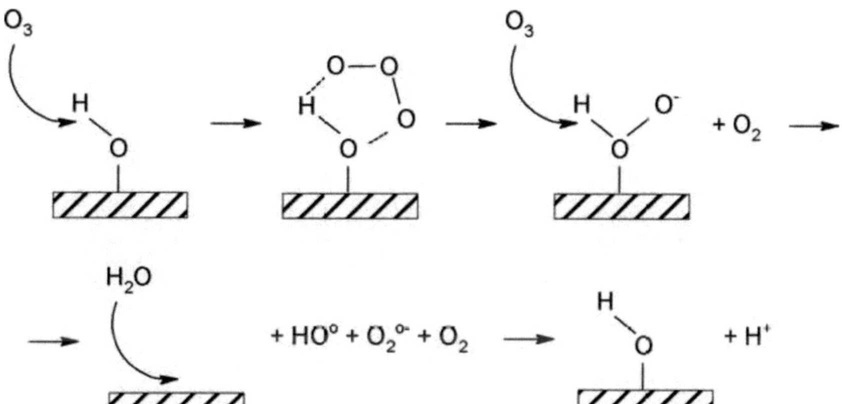

Figure 3.
Mechanism of hydroxyl radical generation in the presence of FeOOH [71].

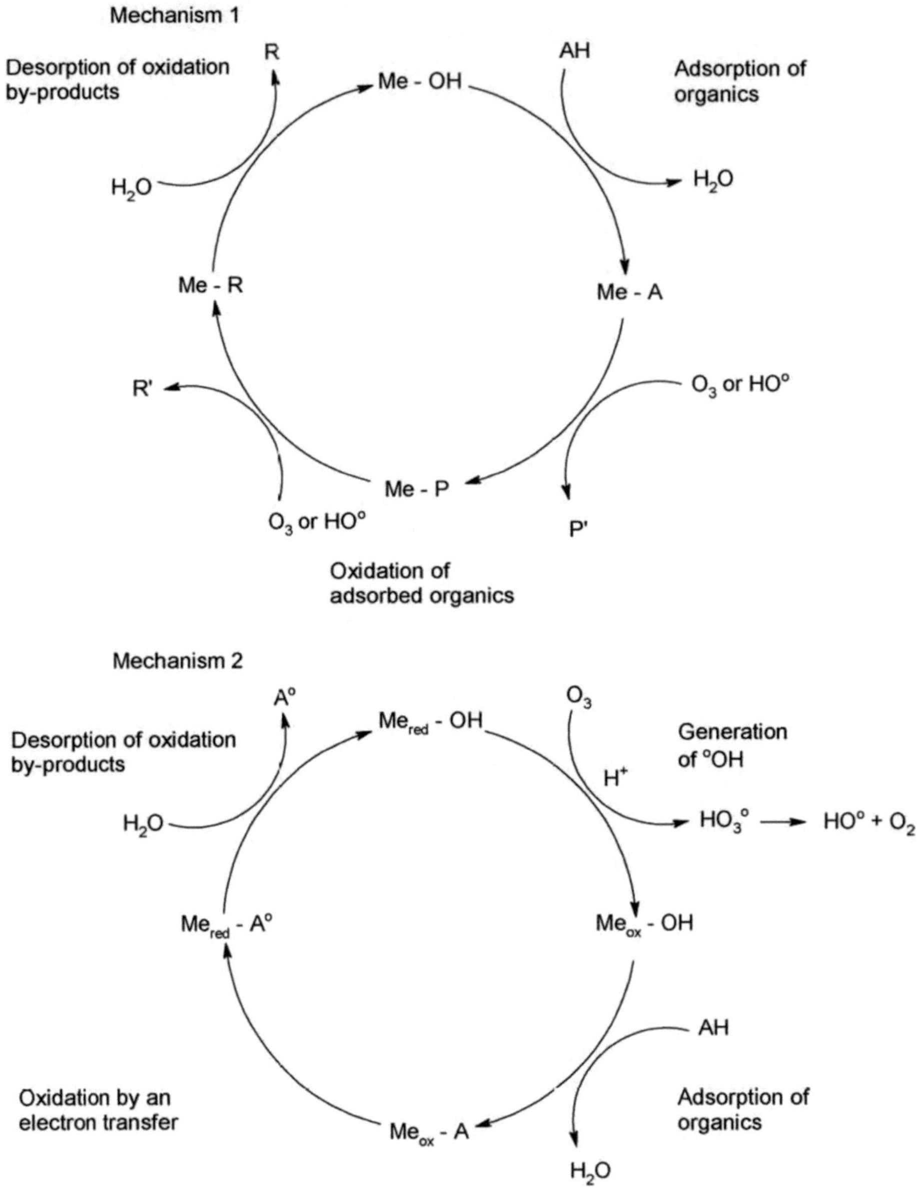

Figure 4.
Mechanisms of catalytic ozone oxidation in the presence of metals on supports (AH (organic acid); P, R (adsorbed primary and final by-products); P, R (primary and final)) [73].

2.2.3 Activated carbon

Activated carbon is also used as a catalyst for catalytic ozone oxidation as a relatively cheap and durable material. The surface properties of activated carbon are different from metals' and cannot coordinate with organic matter in water. So most of the research focused on how ozone decomposes on the surface of activated carbon. In general, there are two catalytic paths for activated carbon: (i) increasing the surface reaction area and (ii) decomposing ozone on the surface to produce •OH radicals.

3. Research on catalytic ozone oxidation in petrochemical secondary effluent

The mechanism described above is part of the research progress on the mechanism of catalytic ozone oxidation in recent years and plays a decisive role in the theoretical development of the process. However, the main research ideas of these studies are to use target pollutants to explore the mechanism of related catalysts. Whether it is possible to maintain high treatment efficiency in the face of actual wastewater with complex water quality is not clear at present; moreover some catalysts have high cost which restrains to implement in practical engineering applications.

In China, ozonation/catalytic ozone oxidation has been widely used in the advanced treatment of petrochemical wastewater. The proportion of applications is estimated to be as high as approximately 64%. Thus, it is necessary to carry out research on the mechanism of catalytic ozone oxidation which is instructive for practical engineering.

The research team of the author is based on a petrochemical wastewater treatment plant (WWTP) which uses catalytic ozone oxidation process as an advanced treatment in the north of China. All the water samples and catalysts in the experiment are obtained from the WWTP, instead of choosing the target pollutants to do in-depth researches, which break through the traditional ways of exploring the mechanism and mean a lot to the practical engineering applications.

In terms of the petrochemical secondary effluent (PSE) treatment, the research on the mechanism of catalytic ozone oxidation and existing problems in the treatment of petrochemical wastewater when using catalytic ozone oxidation process has achieved certain results, which is at the domestic leading level. The research will be generally introduced below.

3.1 Research ideas

The catalysts used in the WWTP are commercial material which mainly comprises alumina-supported copper oxide (ACO). It has been demonstrated that the adsorption of organics on alumina was associated with hydrophilicity of organics [74]. The catalytic activity of dissolved organic matters (DOM) is extremely correlative to their physicochemical properties (e.g., aromaticity, polarity, hydrophobicity-hydrophilicity, acidity-alkaline, molecular weight) [31, 75–77]. Due to the differences of hydrophobicity-hydrophilicity and acidity-alkaline, resin fraction can be an appropriate way to isolate DOM from wastewater [78]. Moreover, the specific effects of ozonation/catalytic ozone oxidation on DOM with different characteristics can be observed from the changes of different DOM fraction. Therefore, the research team innovatively used the resin classification method to divide the DOM in PSE into four components: hydrophobic acids (HoA), hydrophobic bases (HoB), hydrophobic neutrals (HoN), and hydrophilic substances (HI). Herein, HI included hydrophilic acid (HiA), hydrophilic base (HiB), and hydrophilic neutral (HiN). The effect of the three common ozone treatment processes, single ozonation, ozone/H_2O_2 (representing homogeneous catalytic oxidation technology), and catalytic ozone oxidation with a commercial ACO (representing heterogeneous catalytic oxidation technology), on different fractions, was investigated. Then the catalytic ozone oxidation mechanism of the major fractions is systematically elaborated.

3.2 Results and discussion

3.2.1 The characteristic of catalyst

In the study, as the catalyst is commercial and obtained from the practical industrial application, we only investigate the characteristics related to the mechanism proposed. SEM images of commercial catalyst used were presented in **Figure 5**. The catalyst was spherule of 2–4 mm. It possessed mesoporous surface with packed lamellar layer. The structure is very important for the organic pollutant removal. The BET surface area, pore volume, and average size were listed in **Table 2**. The pHpzc of ACO was found to be (pHpzc) of 8.0 ± 0.2. The pH of sample was about 7.5 (pH < pHpzc = 8.0). ACO will be positively charged and adsorb anions from aqueous solution. The surface hydroxyl group density was 2.7×10^{-5} M/m^2. The FTIR spectra of catalyst before and after catalytic ozone oxidation were shown in **Figure 6**. The adsorption peaks at 3450 and 1635 cm^{-1} attributed to the surface hydroxyl groups of catalyst and adsorbed water, respectively. The bands at 1525 and 1385 cm^{-1} were likely to attribute to surface acid functional group of catalyst. After catalytic ozone oxidation, the hydroxyl groups at 3450 cm^{-1} broadened may be formed from adsorbed water by hydrogen binding because the catalyst interacted with ozone.

X-ray photoelectron spectroscopy (XPS) is a surface-sensitive quantitative spectroscopic technique that detects information on the composition and content of various compounds, chemical state, molecular structure, and chemical bonds. As shown in **Figure 7**, obvious O 1s and Al 2p and weak Cu 2p signals are observed in the survey region. The signal corresponding to Al 2p was found at 73.7 eV. The content of Cu was minor. The weak signal related to Cu 2p at around 932 eV could be identified in the narrow region. The signal of O 1s was displayed at 530.3 eV corresponding to the adsorbed hydroxyl.

3.2.2 The effect of ozonation and catalytic ozone oxidation on fraction distribution of the PSE

For the selection study of the optimal wastewater treatment method, Fu et al. compared the reactivity characteristics of DOM in single ozone, ozone/H_2O_2, and

Figure 5.
SEM images for catalyst.

Items	BET surface area (m^2/g)	Pore volume (cm^3/g)	Average pore diameter (nm)	pH_{pzc}	Surface hydroxyl group density (M/m^2)
Average value	169.28	0467	11.0	8.0 ± 0.2	2.7×10^{-5}

Table 2.
Surface characteristics e of commercial catalyst.

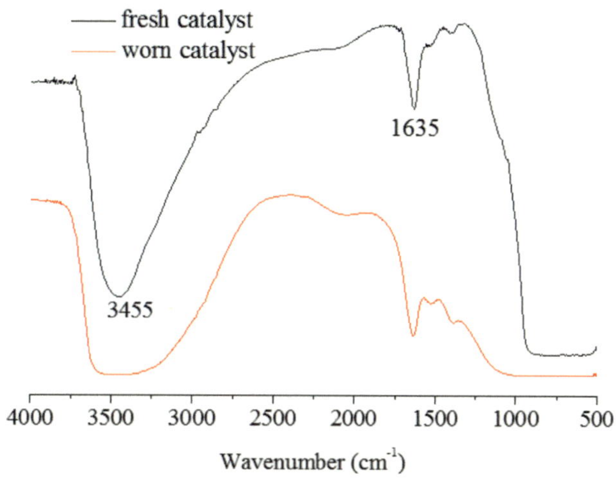

Figure 6.
The FTIR spectra of original and worn catalyst.

ozone/catalyst systems [79]. The highest removal efficiency (42.8%) of TOC was achieved in the ozone/catalyst system, followed by the ozone/H_2O_2 (23.8%) and single ozone (4.7%) systems. This implied that homogeneous catalysis (ozone/H_2O_2) has obvious lower DOM removal rate (~20%) than heterogeneous catalysis when the ozone dosage and contact time were the same treating PSE.

In order to investigate the specific changes of each component, Sun et al. [80] first isolated four fractions (HiA, HiB, HiN, and HI) and operated the ozonation experiment on separate fraction system. They measured the values of DOC and $SUVA_{254}$ of raw water and its fractions before and after ozonation and catalytic ozone oxidation, which present in **Figure 8**. As shown in **Figure 8a**, HI and HoA were the dominated fractions in PSE with relatively high percentage of total DOC. SUA_{254}, the value of UV_{254} divided by itself DOC, has been reported to act as an indicator of active aromatic structure and unsaturated bonds (carbon-carbon and carbon-oxygen) from organic matters [81]. In **Figure 8b**, HoA had the largest $SUVA_{254}$ value among other fractions, indicating HoA had the most containing aromatic and unsaturated bond organic matters, while the $SUVA_{254}$ of HI is the smallest. The results are entirely in accordance with the results of the previous studies [77, 78].

The fraction distribution changed significantly after ozonation and catalytic ozone oxidation. **Figure 8a** manifested that the DOC of HoA remarkably decreased by ozonation. However, the DOC of HI had slight reduction after ozonation. This result indicated that ozonation could be more readily to oxidize HoA than HI. Under the same condition, catalytic ozone oxidation exhibited higher DOC removal than ozonation for all fractions, in particular HI. It is well known that ozone is prone to react with compounds containing electrophilic groups [82]. As can be seen in **Figure 8b**, ozonation decreased significantly the $SUVA_{254}$ of all fractions. On the contrary, catalytic ozone oxidation showed less changes on the reduction of $SUVA_{254}$ than ozonation for all fractions. This suggested UV-absorbing matters are susceptible to electrophilic groups which are easily attacked by ozone. Thus, ozonation reacting easily with HoA is likely to be the highest $SUVA_{254}$.

In a high complex circumstance, Fu et al. evaluated specific changes; six fractions in both raw wastewater and ozonized wastewater instead separate isolated fractions in previous study [80]. Fu et al. specified that HiN (belongs to HI) were one dominant DOM fractions with the highest TOC contents in the effluent of the three ozonation systems, accounting for 24.8–53.6% of the total TOC contents (**Figure 9a**). From the changes in TOC distributions ($TOC_{effluent} - TOC_{influent}$)

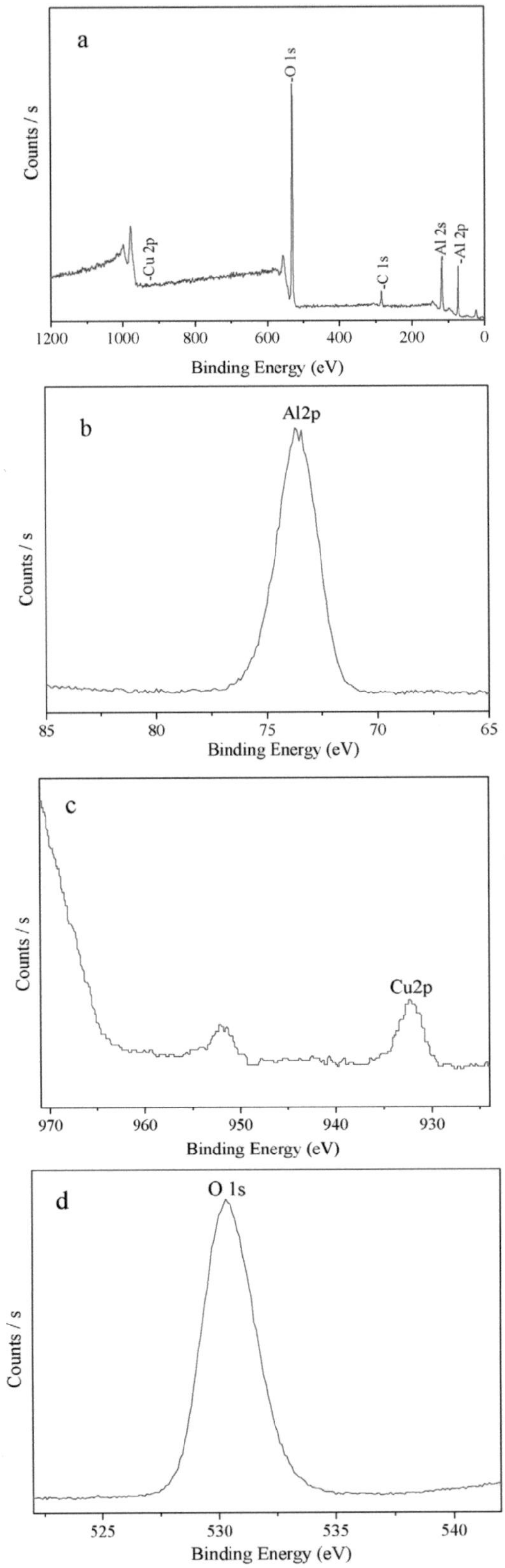

Figure 7.
XPS spectra of catalyst: (a) wide-range survey, (b) Al 2p region, (c) Cu 2p region, and (d) O 1s.

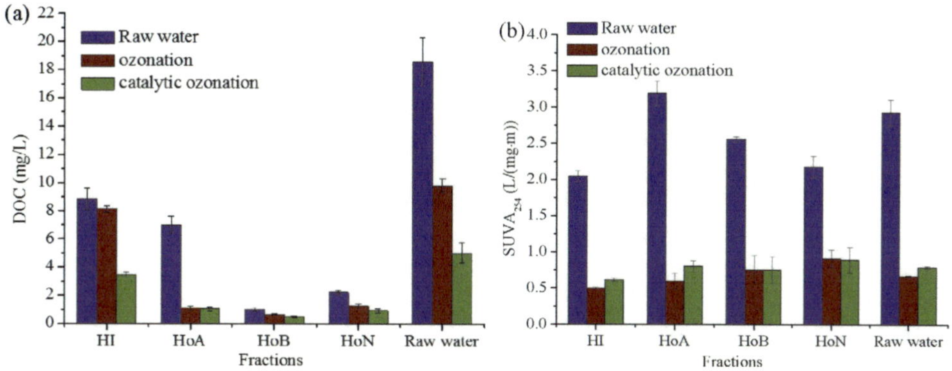

Figure 8.
The DOC (a) and SUVA$_{254}$ (b) of raw water and its fractions before and after ozonation and catalytic ozone oxidation. Experiment conditions: initial concentration, 19.0 ± 2.0 mg/L; initial pH, 7.5; ozone dose, 1.0 mg/min; reaction time, 120 min; catalyst dose, 500 g.

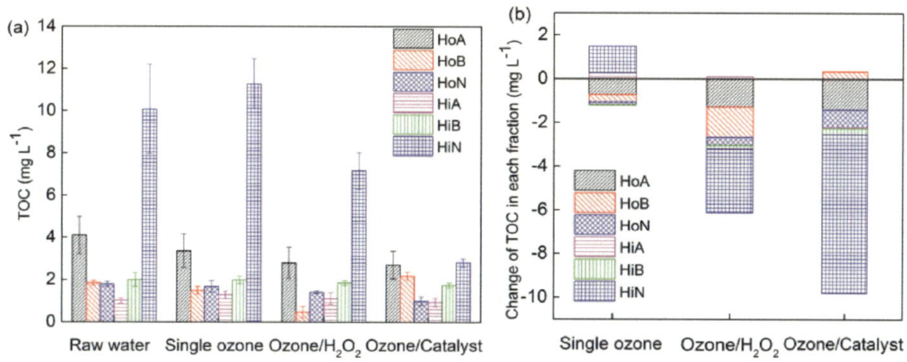

Figure 9.
TOC distribution of the DOM fractions before and after ozonation via single ozone, ozone/H_2O_2, and ozone/catalyst treatment: (a) TOC content and (b) change in TOC content.

(**Figure 9b**), for the single ozone system, the mass balance of TOC content was maintained after single ozone treatment (**Figure 9b**), and the sum of the increased TOC content (contributed by HiA and HiN) was 38.8%, which approximately equaled the sum of the decreased TOC content contributed by the other four fractions (44.8%). In the single ozone system, HoA showed the highest TOC removal, decreasing by 18.1%, whereas HiN exhibited the lowest TOC removal content, increasing by 12.0% in the effluent conversely. Thus, it is reasonable to think that the single ozone system transformed hydrophobic components (e.g., HoB, HoN, and HoA) into hydrophilic compounds (e.g., HiA and HiN). In addition, for ozone/H_2O_2 treatment, HiA was hard to remove, while HoB was easy to remove.

3.2.3 Transformation of DOM fractions during the ozonation process

The specific fluorescence excitation-emission matrix (EEM) spectroscopy volume could be used to rapidly identify the fluorescent compounds present in the DOM and reflect the compositional and structural characteristics of DOM. The ratio of the cumulative volume (Φi) to the TOC (Φi/TOC) of the DOM fractions is further compared (**Figure 10**) [79]. HiA and HiN had the lowest Φi/TOC ratios among other fractions in raw water, indicating that non-fluorescent compounds contributed a large portion of TOC in these two fractions. The FTIR adsorption spectrum of HiA and HiN also suggested that the spectral fine structure and adsorption patterns of the reactive functional groups, such as the stretching vibration of C=O, O—H of the carboxylates, were not as significant as those for the other fractions. The HiN presented the smallest changes in Φi/TOC ratio after single ozone treatment, implying that the portion of non-fluorescent and fluorescent substances did not change much when reacted with single ozone, and HiN may mainly consist of ozone-resistant compounds.

After treatment by the ozone/H_2O_2 and ozone/catalyst systems, the Φi/TOC ratios of HiN in raw water increased significantly (**Figure 10**), showing that compounds with little fluorescence but large TOC contents were removed. Therefore, it could be assumed that the HiN in the raw water may consist of compounds with unsaturated bonds and low molecular weights, which have short conjugation lengths and could readily react with H_2O_2 or catalyst in the presence of ozone [83, 84]. As shown in **Figure 10**, Φi/TOC ratio of HiN did not change substantially in the single ozone system, but increased gradually in the ozone/ H_2O_2 and ozone/catalyst systems. This also demonstrated that HiN was principally responsible for the DOM removal in the ozone/ H_2O_2 system and especially in ozone/catalyst system. Because HiN is the dominant fraction in the sample and was ozone resistant, for the better treatment of petrochemical wastewater, efficient HiN removal is necessary.

3.2.4 The proposed mechanism of catalytic ozone oxidation

1. *HoA removal and transformation*. Sun et al. have proven that HoA is a group of DOM which contains activated aromatic rings or double bonds in their research (**Figure 11**). Those electrophilic groups easily react with ozone molecule [85]. It benefits the contact of HoA and ozone molecule. The study also confirmed that HoA transformed into HI with ozonation. •OH quenching test also proved that ozone molecule and partial •OH reaction are the main pathways.

2. *HI removal*. Higher adsorption rate (22% for HI) on the catalyst of alumina-supported copper oxide (ACO) was observed than that of other studies [74]. The adsorption of catalyst plays a key role during catalytic ozone oxidation. Higher adsorption rate led to the enrichment of organics on ACO, and concentration of HI adsorbed is much higher than that of in the bulking solution. In addition, that adsorbed ozone decomposed into •OH on the surface of

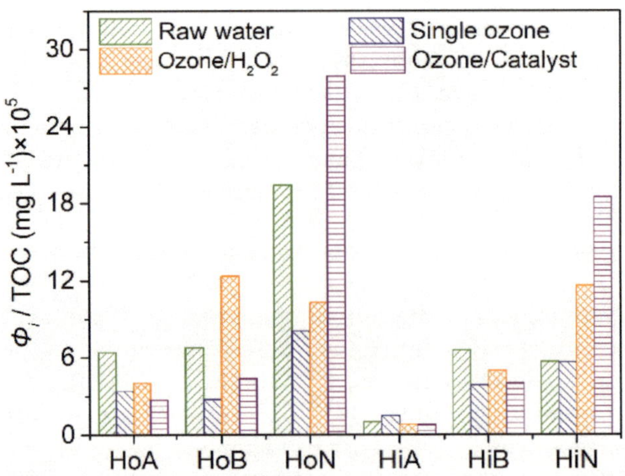

Figure 10.
Ratio of Φi to TOC in the wastewater fractions (Φi/TOC (mg/L) × 10^4 for HiN).

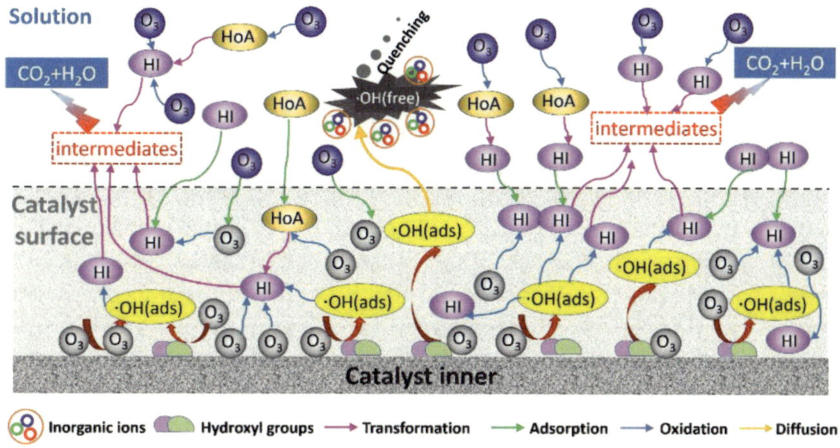

Figure 11.
The proposed reaction mechanism of HoA and HI during catalytic ozone oxidation treating PSE.

catalyst accelerated the mineralization of adsorbed HI, and the reduction of HI on the catalysts' surface would promote the further adsorption of HI from the bulking. Thus, the adsorption role on catalytic ozone oxidation kept a dynamic balance process between ozonation and re-adsorption, triggering a cyclic reaction to reduce the concentration of HI.

3. *OH quenching protection*. Survival time of •OH is just about 10 ms [86]. If the generated •OH on the surface of catalyst are spread into the bulking solution, it can be immediately consumed by high concentration of inorganic salts, such as Cl^- (359.5–380.0 mg/L) and SO_4^{2-} (908.0–955.9 mg/L) with their low affinity for adsorption on the surface of catalyst [87]. Our previous study also found that homogeneous catalysis (H_2O_2/O_3) has obvious lower DOM removal rate (~20%) than heterogeneous catalysis when the ozone dosage and contact time were the same treating PSE [79, 88]. It is proven that the quenching of •OH cannot be ignored during catalytic ozone oxidation of PSE. The adsorption of organic on the surface of catalyst and simultaneous •OH generation in the solid-liquid interface can guarantee the high consumption of •OH reacting with HI, preventing the emission of •OH into the bulking solution. Solid–liquid interface provides dominant reaction site between ozone molecule and •OH with organics. It is the •OH quenching protection.

3.3 Problems on catalytic ozone oxidation in the PSE

3.3.1 The effect of flocs on catalytic ozone oxidation

The toxicity of petrochemical wastewater is higher than that of ordinary urban sewage, which has an adverse effect on the biological treatment system of wastewater, causing the concentration of suspended solids (SS) (mainly organic flocs) in effluent generally beyond 25 mg/L [89]. Generally, SS and colloidal substances in PSE account for 15–20% of the total organic matter in the wastewater. The main components are microbial residues, microbial secretions (such as extracellular polymeric substances, EPSs), soluble microbial metabolites (SMPs), adsorbed proteins, sugars, and other macromolecular organics or inorganic substances such as heavy metals [90]. The presence of flocs in PSE system will have the following adverse effects on the catalytic ozone oxidation unit:

1. An increase in ozone consumption: According to the research, more than 75% of SS in the PSE are small pieces of activated sludge that does not completely separate from the water. The average particle size is about 30–60 μm [91]. Researches discover when catalytic ozone oxidation process is adopted to treat the PSE, the removal of per unit COD will consume more ozone with the increase of SS. The reason is bioflocs in the wastewater can react with O_3, causing the grow of ozone consumption [89]. The research of Zucker et al. [92] shows O_3 has a tendency to react with particulate matter at a high rate even in the first second of the reaction. Our researchers used catalytic ozone oxidation process to treat the PSE. It was found that when the influent SS concentration ranged from 0 to 10 m/L, the depletion of COD was 1.17 gO_3/g COD. When the concentration range of SS rose to 30–35 mg/L, the depletion of COD grew to 2.31 gO_3/g COD. Zhang et al. [90] suggested that when the ozone dosage was less than 10 mg/L, the ozone basically did not react with the flocs in the PSE. Once the dosage was beyond 10 mg/L, ozone can react rapidly with flocs, following the order of dissolved organic matter, outer layer loosely bound EPS (LB-EPS), and inner layer tightly bound EPS (TB-EPS).

2. The flocs covering the catalyst surface hinder the mass transfer of ozone and dissolved organic matters to the catalyst surface, thereby reducing the reaction efficiency. Studies have shown that catalytic ozone oxidation can affect the distribution of the floc particle size in wastewater. It can enhance the flocculation of the flocs, causing the particles to aggregate as larger flocs [92], which are trapped in the catalyst bed. In actual engineering application, if there are no actions being taken to the SS in the PSE before discharging into the unit of catalytic ozone oxidation, the treatment effect will be significantly reduced even within 7 h. Backwashing can effectively improve the situation; however, frequent backwashing strengthens the friction and shear between the catalysts, resulting in the loss of active components of the catalyst, reducing the service life of the catalyst, and invisibly increasing the application cost of the process.

3. Detrimental effects on the removal of characteristic organic matter in the PSE. The research of Zhang et al. [93] indicates that esters, alcohols, olefins, ketones, and nitrogen-containing organic compounds in the PSE can be effectively removed by catalytic ozone oxidation. Twelve main characteristic organic pollutants are detected in the influent, about six kinds of them can be degraded by catalytic ozone oxidation. But the target characteristic pollutants in the effluent can be reduced to three, if the PSE is filtered by 0.45 μm membrane before catalytic ozone oxidation.

3.3.2 The effect of colloidal macromolecular organic matters

Due to the characteristics of small size, light weight, and large specific surface area, the colloidal substance adsorbs a large amount of ions on the surface so that the colloids are mutually repulsive to keep stable state. Colloidal macromolecular substances cannot be settled by their own gravity. So it is necessary to add chemicals in water to make the colloids destabilize and for sedimentation. Studies have shown that colloidal organic matters accounted for about 9%–15% of COD in the PSE [94]. Because of the large molecular weight, such substances will affect the treatment of catalytic ozone oxidation. Adding 25 mg/L poly-aluminum chloride (PAC) coagulant to the PSE to remove the colloidal organic matters before performing catalytic ozone oxidation, compared with the direct use of catalytic ozone oxidation for the PSE, the removal rate of COD can be increased by about 11%. This is because the presence of colloidal macromolecular organics in the PSE affects the rate of ozone decomposition in the initial stage ($t < 30$ s), thereby reducing the production efficiency of hydroxyl radicals (•OH) [95]. Many studies have also shown that in the advanced treatment of sewage, colloidal macromolecular organic matters mainly had effect on the removal of characteristic organic matters during the catalytic ozone oxidation.

3.3.3 The effect on the mass transfer of ozone

Ozone oxidation of organic pollutants in water depends on ozone transferring from the gas phase to the liquid phase, reacting with the target pollutants after diffusion. The mass transfer efficiency of ozone shifting into the liquid phase directly affects the effect of ozone oxidation [96]. The total ozone mass transfer coefficient is mainly influenced by many factors, for example, the degree of gas-liquid two-phase turbulence, the scale and quantity of ozone bubbles, the area of the two-phase contact, and the kinetics of ozonation reaction. The process rate is usually limited by the procedure in which ozone is transferred from the gas phase to the liquid phase [97]. At the same gas flow rate, the smaller bubble size is, the larger phase boundary area will be. Broadening the gas–liquid two-phase boundary area is beneficial not only to increase the total mass transfer rate of the process but also to improve the utilization of ozone [98]. At present, there is still a lack of relevant reports at home and abroad.

4. Conclusions

Flocs and colloidal macromolecular organic matter in the PSE increase the consumption of ozone as well as make the operation unstable. The configuration and operation mode of the catalytic ozone oxidation reactor are not optimized enough, causing the gas-liquid-solid three-phase mass transfer efficiency relatively low. These problems are common problems in the WWTP that are already operating catalytic ozone oxidation processes in China. As a kind of advanced oxidation process, catalytic ozone oxidation has attracted much attention because of its green, high-efficiency and energy-saving characteristics.

People have never stopped exploring its catalytic mechanism; the main research ideas of these studies are to use target pollutants and mixed water to explore the mechanism of related catalysts. Whether it is possible to maintain high treatment efficiency in the face of actual wastewater with complex water quality is not clear at present.

This manuscript first reviews and summarizes the progress and current status of ozonation/catalytic ozone oxidation mechanism research and secondly provides a relatively unique research idea: the DOM in raw water is divided into four fractions, HI, HoA, HoN, and HoB, according to the difference in chemical characteristics. The possible mechanism of ozonation/catalytic ozone oxidation is studied by observing the changes of each fraction. What's more, the common problems that may exist in the actual ozonation/catalytic ozone oxidation process are proposed and summarized. The matching point between mechanism research and practical engineering application is found, which provides direction for future research.

Future research should be carried on further with actual wastewater in the following aspects: (1) Fractional. It is a marvelous research idea to divide the substances in the water into different fractions for the specific research. (2) Microscopic. Future research should focus on microscopic scales, such as phase interface;

it can be combined with orbital theory to discuss the catalytic dynamics in-depth. (3) Whole process. The horizon of researches should be extended to the entire process to explore the effects of the surrounding environment on the catalytic process and mechanism, such as the specific effects of flocs and colloidal molecules on the process. In other words, although we have achieved some results in the study of catalytic ozone oxidation, there is still a lot of work that we need to continue.

Author details

Yu Tan[1], Liya Fu[1], Changyong Wu[1*], Yanan Li[1], Xiumei Sun[1,2] and Yuexi Zhou[1*]

1 State Key Laboratory of Environmental Criteria and Risk Assessment, Chinese Research Academy of Environmental Sciences, Beijing, China

2 College of Water Science, Beijing Normal University, Beijing, China

*Address all correspondence to: changyongwu@126.com and zhouyuexi@263.net

IntechOpen

© 2020 The Author(s). Licensee IntechOpen. Distributed under the terms of the Creative Commons Attribution - NonCommercial 4.0 License (https://creativecommons.org/licenses/by-nc/4.0/), which permits use, distribution and reproduction for non-commercial purposes, provided the original is properly cited.

References

[1] Liao LH. Application of advanced oxidation technology in water supply and wastewater treatment. Chemical Engineering and Equipment. 2018;**08**: 279-281 +259

[2] Beltran FJ, Rivas FJ, Montero-de-Espinosa R. Iron type catalysts for the ozonation of oxalic acid in water. Water Research. 2005;**39**:3553-3564

[3] PIi Y, Ernst M, Schrotter J-CH. Effect of phosphate buffer upon CuO/Al_2O_3 and Cu(II) catalyzed ozonation of oxalic acid solution. Ozone: Science and Engineering. 2003;**25**:393-397

[4] Abd El-Raady AA, Nakajima T. Decomposition of carboxylic acids in water by O_3, O_3/H_2O_2, and O_3/catalyst. Ozone: Science and Engineering. 2005; **27**:11-18

[5] Gracia R, Aragues JL, Ovelleiro JL. Study of the catalytic ozone oxidation of humic substances in water and their ozonation Byproducts. Ozone: Science and Engineering. 1996;**18**:195-208

[6] Piera E, Calpe JC, Brillas E, Domenech X, Peral J. 2,4-Dichlorophenoxyacetic acid degradation by catalyzed ozonation: $TiO_2/UVA/O_3$ and $Fe(II)/UVA/O_3$ systems. Applied Catalysis B: Environmental. 2000;**27**:169-177

[7] Sauleda R, Brillas E. Mineralization of aniline and 4-chlorophenol in acidic solution by ozonation catalyzed with Fe^{2+} and UVA light. Applied Catalysis B: Environmental. 2001;**29**:135-145

[8] Pines DS, Reckhow DA. Effect of dissolved cobalt (II) on the ozonation of oxalic acid. Environmental Science and Technology. 2002;**36**:4046-4051

[9] Beltran FJ, Rivas FJ, Montero-de-Espinosa R. Ozone-enhanced oxidation of oxalic acid in water with cobalt catalysts. 2. Heterogeneous catalytic ozone oxidation. Industrial & Engineering Chemistry Research. 2003; **42**:3210-3217

[10] Logager T, Holeman J, Sehested K, Pedersen T. Oxidation of ferrous ions by ozone in acidic solutions. Inorganic Chemistry. 1992;**31**:3523-3529

[11] Nawrocki J, Kasprzyk-Hordern B. The efficiency and mechanisms of catalytic ozone oxidation. Applied Catalysis B: Environmental. 2010;**99**:27-42

[12] Nawrocki J. Catalytic ozone oxidation in water: Controversies and questions. Discussion paper. Applied Catalysis B: Environmental. 2013; **142-143**:465-471

[13] Beltrán Fernando J, Rivas FJ, Montero-de-Espinosa R. Ozone-enhanced oxidation of oxalic acid in water with cobalt Catalysts.1. Homogeneous catalytic ozone oxidation. Industrial & Engineering Chemistry Research. 2003;**42**(14):3210-3217

[14] Ma J, Graham NJD. Degradation of atrazine by manganese-catalysed ozonation: Influence of humic substances. Water Research. 1999;**33**(3):785-793

[15] Sánchez-Polo M, Rivera-Utrilla J. Ozonation of 1,3,6-naphthalenetrisulfonic acid in presence of heavy metals. Journal of Chemical Technology & Biotechnology. 2004; **79**(8):8

[16] Ma J, Graham NJD. Preliminary investigation of manganese-catalyzed ozonation for the destruction of atrazine. Ozone Science & Engineering. 1997;**19**(3):227-240

[17] Andreozzi R, Casale MSL, Marotta R, et al. N-methyl-p-aminophenol(metol) ozonation in aqueous solution:Kinetics,mechanism

and toxicological characterization of ozonized samples. Water Research. 2000;**34**(18):4419-4429

[18] Cooper C, Burch R. Mesoporous materials for water treatment processes. Water Research. 1999;**33**(18):3689-3694

[19] Kasprzyk-Hordern B, Raczyk-Stanisławiak U, S'wietlik J, Nawrocki J. Catalytic ozone oxidation of natural organic matter on alumina. Applied Catalysis B: Environmental. 2006;**62**: 345-358

[20] Ernst M, Lurot F, Schrotter JC. Catalytic ozone oxidation of refractory organic model compounds in aqueous solution by aluminum oxide. Applied Catalysis B: Environmental. 2004;**47**(1): 15-25

[21] Yunrui Z, Wanpeng Z, Fudong L, et al. Catalytic activity of Ru/$A_{12}O_3$ for ozonation of dimethyl phthalate in aqueous solution. Chemosphere. 2007; **66**(1):145-150

[22] Ni CH, Chen JN. Heterogeneous catalytic ozone oxidation of 2-chlorophenol aqueous solution with alumina as a catalyst. Water Science & Technology. 2001;**43**(2):213

[23] Beltrán FJ, Rivas FJ, Montero-De-Espinosa R. Catalytic ozone oxidation of oxalic acid in an aqueous TiO slurry reactor. Applied Catalysis B: Environmental. 2002;**39**(3):221-231

[24] Rosal R, Rodríguez A, Gonzalo MS, et al. Catalytic ozone oxidation of naproxen and carbamazepine on titanium dioxide. Applied Catalysis B: Environmental. 2008;**84**(1):48-57

[25] Rosal R, Gonzalo MS, Rodríguez A, et al. Ozonation of clofibric acid catalyzed by titanium dioxide. Journal of Hazardous Materials. 2009;**169**(1): 411-418

[26] Rosal R, Gonzalo MS, Boltes K, et al. Identification of intermediates and assessment of ecotoxicity in the oxidation products generated during the ozonation of clofibric acid. Journal of Hazardous Materials. 2009;**172**(2): 1061-1068

[27] Zhang T, Lu J, Ma J, et al. Fluorescence spectroscopic characterization of DOM fractions isolated from a filtered river water after ozonation and catalytic ozone oxidation. Chemosphere. 2008;**71**(5):911-921

[28] Park JS, Choi H, Ahn KH. The reaction mechanism of catalytic oxidation with hydrogen peroxide and ozone in aqueous solution. Water Science & Technology. 2003;**47**(1):179

[29] Park JS, Choi H, Ahn KH, et al. Removal mechanism of natural organic matter and organic acid by ozone in the presence of goethite. Ozone Science & Engineering. 2004;**26**(2):141-151

[30] Park JS, Choi H, Cho J. Kinetic decomposition of ozone and para-chlorobenzoic acid (CBA) during catalytic ozone oxidation. Water Research. 2004;**38**(9):2285-2292

[31] Zhang T, Lu J, Ma J, et al. Comparative study of ozonation and synthetic goethite-catalyzed ozonation of individual NOM fractions isolated and fractionated from a filtered river water. Water Research. 2008;**42**(6): 1563-1570

[32] Trapido M, Veressinina Y, Munter R, et al. Catalytic ozone oxidation of m-dinitrobenzene. Ozone Science & Engineering. 2005;**27**(5): 359-363

[33] Andreozzi R, Insola A, Caprio V, et al. The use of manganese dioxide as a heterogeneous catalyst for oxalic acid ozonation in aqueous solution. Applied Catalysis A: General. 1996;**138**(1):75-81

[34] Muruganandham M, Wu JJ. Synthesis, characterization and catalytic activity of easily recyclable zinc oxide nanobundles. Applied Catalysis B: Environmental. 2008;**80**(1):32-41

[35] Gracia R, Cortés S, Sarasa J, et al. Heterogeneous catalytic ozone oxidation with supported titanium dioxide in model and natural waters. Ozone Science & Engineering. 2000;**22**(5): 461-471

[36] Gracia R, Cortés S, Sarasa J, et al. Catalytic ozone oxidation with supported titanium dioxide. The stability of catalyst in water. Ozone Science & Engineering. 2000;**22**(2): 185-193

[37] Gracia R, Cortes S, Sarasa J, et al. TiO_2-catalysed ozonation of raw Ebro river water. Water Research. 2000; **34**(5):1525-1532

[38] Volk C, Roche P, Joret JC, et al. Comparison of the effect of ozone, ozone-hydrogen peroxide system and catalytic ozone on the biodegradable organic matter of a fulvic acid solution. Water Research. 1997;**31**(3): 650-656

[39] Beltrán FJ, Rivas FJ, Montero-De-Espinosa R. A TiO_2/Al_2O_3 catalyst to improve the ozonation of oxalic acid in water. Applied Catalysis B: Environmental. 2004;**47**(2):101-109

[40] Ma J, Sui MH, Chen ZL, et al. Degradation of refractory organic pollutants by catalytic ozone oxidation —Activated carbon and Mn-loaded activated carbon as catalysts. Ozone Science & Engineering. 2004;**26**(1):3-10

[41] Álvarez PM, Beltrán FJ, Pocostales JP, et al. Preparation and structural characterization of Co/Al_2O_3 catalysts for the ozonation of pyruvic acid. Applied Catalysis B: Environmental. 2007;**72**(3):322-330

[42] Chen L, Qi F, Xu B, et al. The efficiency and mechanism of γ-alumina catalytic ozone oxidation of 2-methylisoborneol in drinking water. Water Science & Technology: Water Supply. 2006; **6**(3):43

[43] Qi F, Xu B, Chen Z, et al. Catalytic ozone oxidation for degradation of 2,4,6-trichloroanisole in drinking water in the presence of gamma-AlOOH. Water Environment Research. 2009; **81**(6):592

[44] Dong Y, Yang H, He K, et al. β-MnO_2 nanowires: A novel ozonation catalyst for water treatment. Applied Catalysis B: Environmental. 2009;**85**(3): 155-161

[45] Faria PCC, Monteiro DCM, Órfão JJM, et al. Cerium, manganese and cobalt oxides as catalysts for the ozonation of selected organic compounds. Chemosphere. 2009;**74**(6): 818-824

[46] Faria PCC, Órfão JJM, Pereira MFR. Activated carbon and ceria catalysts applied to the catalytic ozone oxidation of dyes and textile effluents. Applied Catalysis B: Environmental. 2009;**88**(3): 341-350

[47] Qu J, Li H, Liu H, et al. Ozonation of alachlor catalyzed by Cu/Al_2O_3 in water. Catalysis Today. 2004;**89**(3):291-296

[48] Udrea I, Bradu C. Ozonation of substituted phenols in aqueous solutions over CuO-Al_2O_3 catalyst. Ozone Science & Engineering. 2003; **25**(4):335-343

[49] Xu ZZ, Chen ZL, Joll C, et al. Catalytic efficiency and stability of cobalt hydroxide for decomposition of ozone and-chloronitrobenzene in water. Catalysis Communications. 2009;**10**(8): 1221-1225

[50] Jie Z, Lee KH, Cui L, et al. Degradation of methylene blue in aqueous solution by ozone-based processes. Journal of Industrial & Engineering Chemistry. 2009;**15**(2): 185-189

[51] Avramescu SM, Bradu C, Udrea I, et al. Degradation of oxalic acid from aqueous solutions by ozonation in presence of Ni/Al_2O_3 catalysts. Catalysis Communications. 2008;**9**(14): 2386-2391

[52] Qin W, Li X, Qi J. Experimental and theoretical investigation of the catalytic ozone oxidation on the surface of NiO−CuO nanoparticles. Langmuir: The ACS Journal of Surfaces & Colloids. 2009;**25**(14):8001-8011

[53] Yang L, Hu C, Nie Y, et al. Catalytic ozone oxidation of selected pharmaceuticals over mesoporous alumina-supported manganese oxide. Environmental Science & Technology. 2009;**43**(7):2525-2529

[54] Moussavi G, Mahmoudi M. Degradation and biodegradability improvement of the reactive red 198 azo dye using catalytic ozone oxidation with MgO nanocrystals. Chemical Engineering Journal. 2009;**152**(1):1-7

[55] Hu C, Xing S, Jiuhui Qu A, et al. Catalytic ozone oxidation of herbicide 2,4-D over cobalt oxide supported on mesoporous zirconia. Journal of Physical Chemistry C. 2008;**112**(15):5978-5983

[56] Sui M, Sheng L, Lu K, et al. FeOOH catalytic ozone oxidation of oxalic acid and the effect of phosphate binding on its catalytic activity. Applied Catalysis B: Environmental. 2010;**96**(1):94-100

[57] Rui CM, Quinta-Ferreira RM. Catalytic ozone oxidation of phenolic acids over a Mn-Ce-O catalyst. Applied Catalysis B: Environmental. 2009;**90**(1): 268-277

[58] Tong S-P, Liu W-P, Leng W-H, Zhang Q-Q. Characteristics of MnO_2 catalytic ozone oxidation of sulfosalicylic acid and propionic acid in water. Chemosphere. 2003;**50**:1359-1364

[59] Ma J, Sui M, Zhang T, et al. Effect of pH on MnO_x/GAC catalyzed ozonation for degradation of nitrobenzene. Water Research. 2005;**39**(5):779-786

[60] Kasprzyk B, Nawrocki J. Preliminary results on ozonation enhancement by a perfluorinated bonded alumina phase. Ozone: Science and Engineering. 2002;**24**:63-68

[61] Kasprzyk-Hordern B, Nawrocki J. The feasibility of using a perfluorinated bonded alumina phase in the ozonation process. Ozone: Science and Engineering. 2003;**25**:185-194

[62] Kasprzyk-Hordern B, Andrzejewski P, Dąbrowska A, Czaczyk K, Nawrocki J. MTBE, DIPE, ETBE and TAME degradation in water using perfluorinated phases as catalysts for ozonation process. Applied Catalysis B: Environmental. 2004;**51**:51-66

[63] Han F, Kambala VSR, Srinivasan M, Rajarathnam D, Naidu R. Tailored titanium dioxide photocatalysts for the degradation of organic dyes in wastewater treatment: A review. Applied Catalysis A: General. 2009;**359**:25-40

[64] Gaya UI, Abdullah AH. Heterogeneous photocatalytic degradation of organic contaminants over titanium dioxide: A review of fundamentals, progress and problems. Journal of Photochemistry and Photobiology C: Photochemistry Reviews. 2008;**9**:1-12

[65] Chatterjee D, Dasgupta S. Visible light induced photocatalytic degradation of organic pollutants. Journal of Photochemistry and Photobiology C: Photochemistry Reviews. 2005;**6**:186-205

[66] Suty H, de Traversay C, Cost M. Applications of advanced oxidation processes: Present and future. Water Science and Technology. 2004;**49**: 227-233

[67] Bhatkhande DS, Pangarkar VG. Photocatalytic degradation for environmental applications-a review Journal of Chemical Technology and Biotechnology. 2002;**77**:102-116

[68] Chong MN, Jin B, Chow Christopher WK, Saint C. Recent developments in photocatalytic water treatment technology: A review. Water Research. 2010;**44**(10): 2997-3027

[69] Yang Y, Ma J, Qin Q, Zhai X. Degradation of nitrobenzene by nano-TiO_2 catalyzed ozonation. Journal of Molecular Catalysis A: Chemical. 2007; **267**:41-48

[70] Ye M, Chen Z, Liu X, Ben Y, Shen J. Ozone enhanced activity of aqueous titanium dioxide suspensions for photodegradation of 4-chloronitrobenzene. Journal of Hazardous Materials. 2009;**167**: 1021-1027

[71] Zhang T, Ma J. Catalytic ozone oxidation of trace nitrobenzene in water with synthetic goethite. Journal of Molecular Catalysis A: Chemical. 2008; **279**:82-89

[72] Ma J, Zhang T, Chen ZL. The mechanism of ozone decomposition and oxidation of trace nitrobenzene catalyzed by ferric hydroxide in water. Environmental Science. 2005;**26**(2): 78-82

[73] Legube B, Karpel Vel Leitner N. Catalytic ozone oxidation: A promising advanced oxidation technology for water treatment. Catalysis Today. 1999; **53**(1):61-72

[74] Ikhlaq A, Brown DR, Kasprzyk-Hordern B. Catalytic ozone oxidation for the removal of organic contaminants in water on alumina. Applied Catalysis B: Environmental. 2015;**165**:408-418

[75] Qi F, Chen ZL, Xu BB, Shen JM, Ma J, AnnaHeitz CJ. Influence of surface texture and acid-base properties on ozone decomposition catalyzed by aluminum (hydroxyl) oxides. Applied Catalysis B: Environmental. 2008;**84**: 684-690

[76] Qian FY, Sun XB, Liu YD. Effect of ozone on removal of dissolved organic matter and its biodegradability and adsorbability in biotreated textile effluents. Ozone Science & Engineering. 2013;**35**:7-15

[77] Li CM, Wang DH, Xu X, Wang ZJ. Formation of known and unknown disinfection by-products from natural organic matter fractions during chlorination, chloramination, and ozonation. Science of the Total Environment. 2017;**587-588**:177-184

[78] Gong JL, Liu YD, Sun XB. O_3 and UV/O_3 oxidation of organic constituents of biotreated municipal wastewater. Water Research. 2008;**42**:1238-1244

[79] Fu LY, Wu CY, Zhou YX, Zuo JE, Song GQ, Tan Y. Ozonation reactivity characteristics of dissolved organic matter in secondary petrochemical wastewater by single ozone, ozone/ H_2O_2, and ozone/catalyst. Chemosphere. 2019;**233**:34-43

[80] Sun XM, Wu CY, Zhou YX, Han W. Using DOM fraction method to investigate the mechanism of catalytic ozonation for real wastewater. Chemical Engineering Journal. 2019; **369**:100-108

[81] Ratpukdi T, Siripattanakul S, Khan E. Mineralization and biodegradability enhancement of

natural organic matter by ozone—VUV in comparison with ozone, VUV, ozone—UV, and UV: Effects of pH and ozone dose. Water Research. 2010;**44**: 3531-3542

[82] Hoigné J, Bader H. Rate constants of reactions of ozone with organic and inorganic compounds in water—I: Non-dissociating organic compounds. Water Research. 1983;**17**:173-183

[83] Baker A. Fluorescence excitation–emission matrix characterization of some sewage-impacted rivers. Environmental Science & Technology. 2001;**35**:948-953

[84] Chen W, Westerhoff P, Leenheer JA, Booksh K. Fluorescence excitation–emission matrix regional integration to quantify spectra for dissolved organic matter. Environmental Science & Technology. 2003;**37**:5701-5710

[85] Van Sonntag C, Van Gunten U. Chemistry of Ozone in Water and Wastewater Treatment. London, UK: IWA Press; 2012

[86] Wu JF, Su TM, Jiang YX, Xie XL, Qin ZZ, Ji HB. In situ DRIFTS study of O_3 adsorption on CaO, γ-Al_2O_3, CuO, γ-Fe_2O_3 and ZnO at room temperature for the catalytic ozone oxidation of cinnamaldehyde. Applied Surface Science. 2017;**412**:290-305

[87] Petre AL, Carbajo JB, Rosal R, Garcia-Calvo E, Perdigón-Melón JA. CuO/SBA-15 catalyst for the catalytic ozone oxidation of mesoxalic and oxalic acids: Water matrix effects. Chemical Engineering Journal. 2013;**225**:164-173

[88] Fu LY. The study of the operation and mechanism of an integrated ozone and biological aerated filter (BAF) advanced treatment for petrochemical wastewater [PhD dissertation]. Beijing, China: Tsinghua University; 2017

[89] Wu C, Li Y, Zhou Y, Li Z. Upgrading the Chinese biggest petrochemical wastewater treatment plant: Technologies research and full scale application. Science of the Total Environment. 2018;**633**:189-197

[90] Zhang SY, Wu CY, Zhou YX. Effects of bioflocs on advanced treatment of petrochemical secondary effluent by ozone catalytic oxidation. Proceedings of the Annual Meeting of the Chinese Society of Environmental Sciences. 2017:2869-2877

[91] Wu C, Zhou Y, et al. Pretreatment of petrochemical secondary effluent by micro-flocculation and dynasand filtration: Performance and DOM removal characteristics. Water Air & Soil Pollution. 2016;**227**(11):415

[92] Zucker I, Lester Y, Avisar D, et al. Influence of wastewater particles on ozone degradation of trace organic contaminants. Environmental Science and Technology. 2015;**9**:301-308

[93] Zhang S, Wu C, Zhou Y, et al. Effect of wastewater particles on catalytic ozone oxidation in the advanced treatment of petrochemical secondary effluent. Chemical Engineering Journal. 2018;**345**:280-289

[94] Rickert DA, Hunter JV. Colloidal matter in wastewaters and secondary effluents. Journal of Water Pollution Control Federation. 1972;**44**(1):134-139

[95] Wert EC, Gonzales S, Dong MM, et al. Evaluation of enhanced coagulation pretreatment to improve ozone oxidation efficiency in wastewater. Water Research. 2011; **45**(16):5191-5199

[96] Qin YJ, Qin WZ, Yang PF. Advances in enhanced ozone mass transfer. Journal of Process Engineering. 2017; **17**(2):420-426

[97] Li X, Lin C, Jiang Y. Effect of microporous distributor on mass transfer in photocatalytic ozonation reactor. Journal of Environmental Engineering. 2011;**5**(5):1101-1105

[98] Chu JY, Wu CD, Chen WJ. Ozone technology and its application. Beijing: Chemical Industry Press; 2002

Chapter 8

Scale-Up and Optimization for Slurry Photoreactors

Gianluca Li Puma, Fiderman Machuca-Martínez, Miguel Mueses, José Colina-Márquez and Ciro Bustillo-Lecompte

Abstract

Several aspects of scale-up and optimization of heterogeneous photocatalytic reactors are detailed in the following chapter. The relevance of the dimensionless numbers, the critical factors of the photoreactor design, and the optimization methods are explored from an engineering point of view. The apparent optical thickness is the most crucial dimensionless parameter for photoreactor design and optimization; therefore, the study will be more focused on this topic. Moreover, a real case for a commercial application of the solar photocatalysis will be presented in this chapter. This full-scale plant is currently operating as an industrial wastewater treatment plant in a flexographic company located in Cali, Colombia.

Keywords: scale-up, heterogeneous photocatalysis, full-scale, advanced wastewater treatment

1. Introduction

For scaling-up heterogeneous photocatalytic reactors, it is necessary to consider the following key factors: kinetic law, UV radiation source, photoreactor type, catalyst type (suspended or fixed), and mass transfer phenomena. The performance of the photocatalytic reactor depends directly on the number of photons absorbed by the catalyst surface; and this is related to the geometry of the photoreactor, UV photons source, and how the catalyst interacts with the chemical species. Although the photoreactor models consider practically all these aspects, the use of dimensionless parameters is highly recommended for scaling-up purposes. The four dimensionless numbers needed for the scaling-up process are as follows [1]:

- Reynolds number (N_{Re}), which suggests the type of fluid regime to be considered for the reactor model.

- Damkohler number (N_{Da}), which expresses the ratio of the overall reaction rate, calculated at the inlet concentration and at maximum photon flux, to the maximum input mass flow rate of the reactant model (laminar or turbulent flow) to be included in the model. The values estimated for this dimensionless number allow identifying which can be the slower stage of the overall reaction rate.

 IntechOpen

- The optical thickness of the photoreactor (τ), corresponding to the degree of opacity of the photoreactor.

- The scattering albedo of the photocatalyst (ω), which depends on the optical properties of the catalyst.

The last two dimensionless numbers (optical thickness and scattering albedo) are the most relevant for scaling up photocatalytic reactors. Both help to describe the dynamics of the UV photon absorption and determine the level of difficulty of the scaling-up process. For example, some thin-film slurry (TFS) photoreactors (**Figure 1**) can exhibit very low values of scattering albedo ($\omega < 0.3$); thus, the rate of photon absorption (LVRPA) can be estimated by neglecting the scattering effect safely, and the scaling-up process can be more straightforward for this case; whereas, for TFS with higher scattering albedos or "geometrically thick" slurry photoreactors, such as compound parabolic collectors (CPC) and tubular reactors, the LVRPA needs to be modeled with more complex mathematical expressions due to the scattering effect cannot be neglected [1].

Although TFS reactors with low scattering albedos are easier to scale-up (regarding photonic effects calculations), they can result not so practical when they need to handle larger volumes of reaction, especially if the dimensionless numbers need to remain constant in the process of scaling-up. Several aspects of scale-up and optimization of heterogeneous photocatalytic reactors are detailed in the following sections of this chapter.

2. Scale-up of heterogeneous photocatalytic reactors

The dimensions of a full-scale TFS reactor can be challenging to handle if, for example, a large plate of several meters of length and width needs to be located in a reduced space. This is not the case for CPC or tubular photoreactors, which can be set up as modular units for larger reaction volumes. This advantage is more relevant when the photon source is the UV solar radiation since the photocatalytic reactors must be located outdoors, and the TFS reactors must have additional protection for weather phenomena. The solar CPC photoreactors, instead, are more robust, and there are several large-scale applications as it has been reported in the literature [1–5].

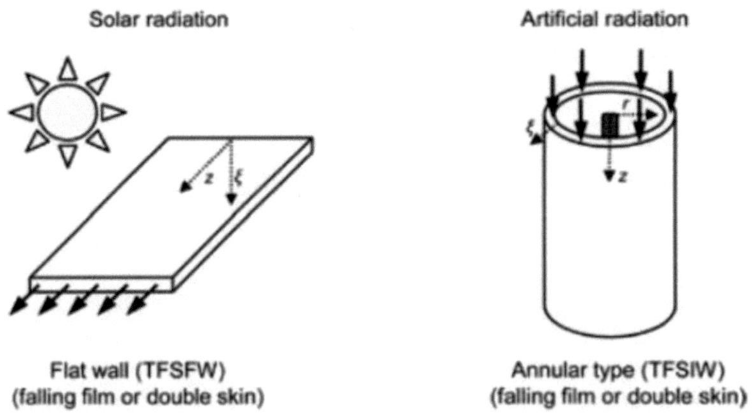

Figure 1.
Thin-film slurry (TFS) photoreactors. © 2005 Elsevier Science Ltd. Originally published in Li Puma [1] under CC BY-NC-ND 4.0 license.

Nonetheless, other "geometrically thick" reactors can be considered as feasible options to be scaled-up, regarding the cost-benefit ratio.

The optical thickness, which depends on the catalyst load and the reactor thickness, may be manipulated in order to maximize the photon absorption by the catalyst. Consequently, it is necessary to observe how the LVRPA behaves with different values of optical thickness, considering that the type of reactor is relevant to the behavior of this parameter.

Figure 2 shows how the conversion of an ideal substrate varies with the optical thickness and the type of reactor: falling film laminar flow (FFLF), plug flow (PF), and slit flow (SF) reactors. It is also clear that the lower scattering albedos allow higher conversions of this ideal substrate.

The "optically thick" reactors ($\tau > 1.0$) exhibit lower conversions due to a high scattering effect although there are also high photon absorption rates due to the higher amounts of catalyst. For this kind of photoreactors, the optimization of the apparent optical thickness is strongly recommended in order to use the minimum amount of catalyst that ensures satisfactory performance of the photocatalytic reactor.

Other dimensionless numbers or parameters can be used. The dimensionless intensity of radiation at the entrance window of the reactor and the dimensionless LVRPA may be useful as well. The following expressions apply for TFS reactors ($\omega < 0.3$) [1]:

$$I^*_{\xi^*=0,z^*} = \frac{I_{\xi^*=0,z^*}}{\left(I^{max}_{\xi^*=0}\right)_{\omega=0}} \tag{1}$$

$$LVRPA^* = I^*_{\xi^*=0,z^*} \exp\left(-\tau_{app}\xi^*\right) \tag{2}$$

where $I^*_{\xi^*=0,z^*}$ is the dimensionless intensity at the entrance window ($\xi^* = 0$); $I_{\xi^*=0,z^*}$ is the actual intensity estimated at the entrance window; and $\left(I^{max}_{\xi^*=0}\right)_{\omega=0}$ is the maximum intensity of the radiation estimated in the absence of scattering effects.

The dimensionless LVRPA is easily estimated from the dimensionless intensity at the entrance window, the apparent optical thickness (τ), and the dimensionless

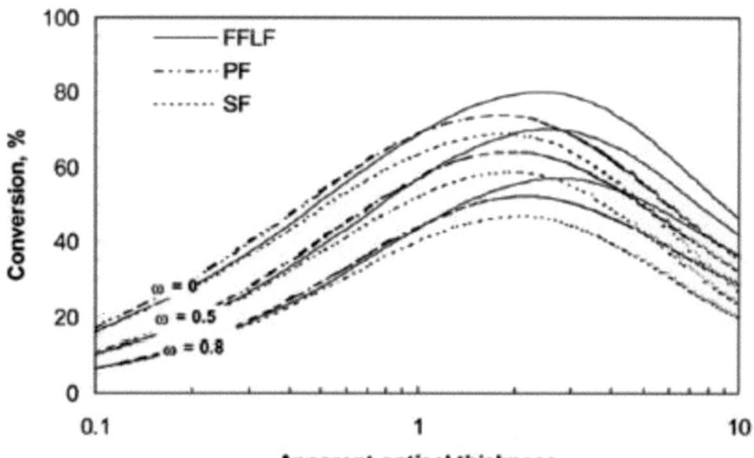

Figure 2.
Model simulations for the conversion of a substrate as a function of the optical thickness and scattering albedo for different idealized flow conditions. © 2005 Elsevier Science Ltd. Originally published in Li Puma [1] under CC BY-NC-ND 4.0 license.

radial coordinate (ξ^*). As mentioned earlier, the problem with those expressions is that they are limited to cases where the scattering albedo is low. However, that is not the case for the commercial TiO_2 used in heterogeneous photocatalytic reactions, which exhibits higher values of ω [1].

The dimensionless intensity can be used when a photocatalytic reactor with artificial UV radiation source is going to be scaled-up. Nevertheless, this expression cannot be applied for solar photoreactors because of the variability of the solar incident radiation. For this case, the LVRPA must be calculated based on the solar radiation emission models, considering the location latitude, time, and date and the clearness factor (K_C) [2]. This K_C parameter depends on the number of clouds at the time of the photocatalytic reaction, which varies on the weather conditions [3].

3. Optimization methods

The optimization for photocatalytic reactors is usually intended for finding the operating conditions that ensure the best performance regarding the degradation of a specific substrate by photochemical oxidation. For the scaling-up process, this is a necessary step for making the heterogeneous photocatalysis a profitable and competitive technology.

The optimization can be carried out from the experimental data of pilot-scale photoreactors by using a simple empirical model. This model is built from a polynomial expression that involves the most relevant operating variables that affect the interest variable (e.g. relative degradation of the substrate).

Figure 3 shows an example where the surface response methodology was used for optimizing the decolorization of methylene blue (MB) with two bench-scale photoreactors [6].

In the reported study by Arias et al. [6], the tubular reactor corresponds to a 38 mm-ID borosilicate glass tube of 30 cm of length (**Figure 4a**), whereas the CPC reactor consists of the same glass tube but with a reflective collector that redistributes the reflected radiation (**Figure 4b**).

For each case, the slurry was recirculated during 30 min under darkness conditions, in order to achieve the adsorption equilibrium of the dye on the catalyst surface. In previous results, the adsorption removal of the dye was around 10–15%. It was possible to build the polynomial models from results of a full-composite experimental design, where the TiO_2 load and the initial pH of the slurry were selected as the controllable factors and the relative decolorization was the response variable. The optimization was made by using the first-derivative method for the

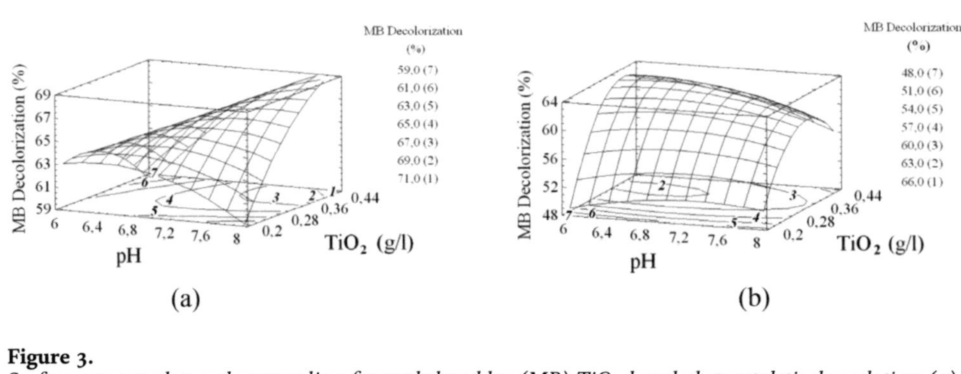

Figure 3.
Surface response plots and contour lines for methylene blue (MB) TiO_2-based photocatalytic degradation: (a) CPC reactor and (b) tubular reactor. © 2005 Walter de Gruyter & Co. Originally published in Arias et al. [6] under CC BY-NC-ND 4.0 license.

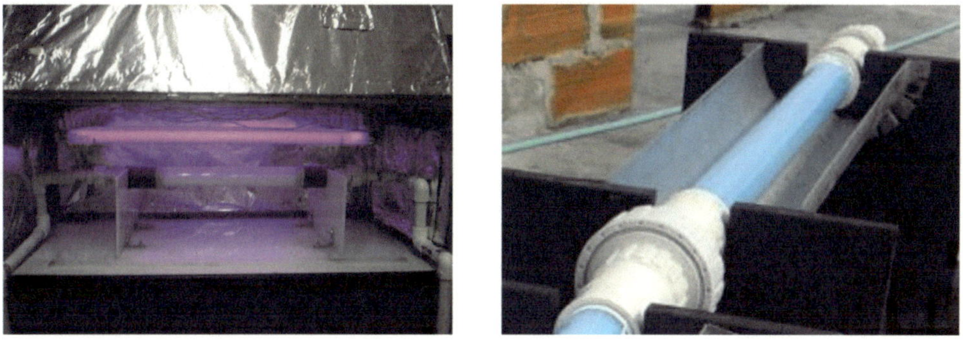

(a) (b)

Figure 4.
Bench-scale photoreactors used for MB decolorization; (a) tubular and (b) CPC.

obtained empirical expressions, and the optimal catalyst load and pH were found for each photoreactor [6].

This method allows finding optimums without a significant mathematical effort; however, its results only are scalable within the experimental conditions used for the pilot or bench-scale tests. Besides, the effects of mass transfer and the photon absorption effects are not considered in this empirical model. This leads to using of photoreactors of the same dimensional and hydrodynamic features than that used for the experimental tests. Therefore, this optimization method is limited because of its empirical nature, and it is necessary to consider a method that involves dimensionless numbers, more suitable for scaling-up.

The other method consists of optimizing the mathematical expressions of the photoreactor model. The first derivative (as optimization tool) might not be practical for finding the maximums or minimums of these mathematical expressions due to their intrinsic complexity, especially for "geometrically thick" photoreactors. Therefore, it is more recommendable to evaluate the property to be optimized in a wide range of the dependent variables. Despite the demanding calculation time of this type of optimization, this strategy is more applicable for the scaling-up process since the use of dimensionless numbers in the optimization ensures the similarity between scales.

The most suitable dimensionless parameter being used in the optimization is τ. Generally, the variable to be optimized is the volumetric rate of photon absorption (VRPA). However, the degradation of a specific substrate can be further used. **Figure 2** shows the behavior of a hypothetical conversion vs. the apparent optical thickness and the plots that exhibit an optimum.

From the LVRPA expression and supposing a constant UV radiation flux (I_0) of 30 W/m^2, the following plots can be obtained in order to optimize the rate of photon absorption in the CPC or tubular reactors (**Figure 5**).

The plots shown in **Figure 6** describe the behavior of the VRPA/H with tubular and CPC reactors of 32 mm ID and AEROXIDE® P-25 as the catalyst. From the previous definitions, these reactors are considered "optically thick," and the LVRPA must be calculated using the six-flux model (SFM) approach. It is important to note that these plots are only applicable to the conditions of radiation and the tube diameter specified [7]. Although for scaling-up modular photoreactors, such as the CPC or tubular, the tube diameter can remain constant in the process. In the case of different diameter (or different reactor thickness), τ is the most suitable parameter to be involved in the optimization. **Figure 6** shows a similar trend to the one shown in **Figure 2**. The presence of optimums in each case is because of the scattering increases with the optical thickness, generally associated to the high catalyst loads that do not permit a satisfactory light penetration inside the reactor bulk.

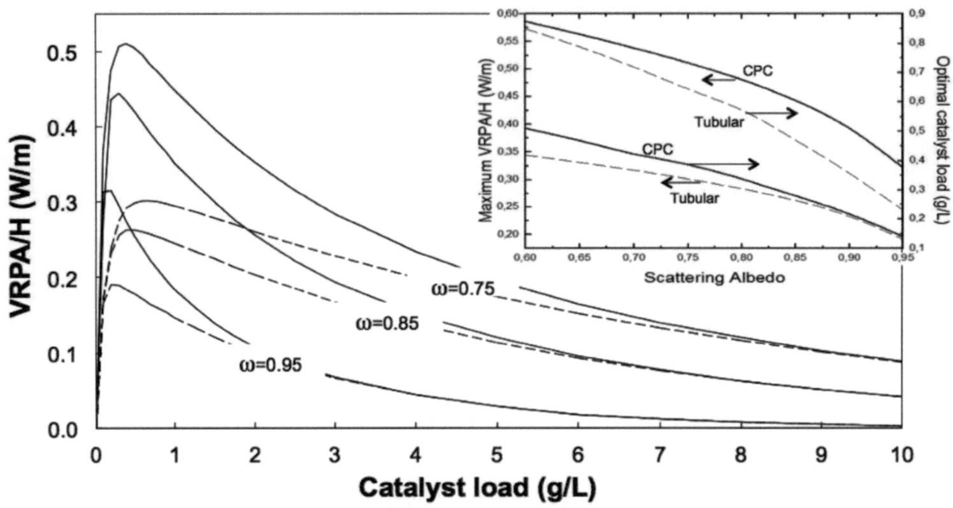

Figure 5.
Volumetric rate of photon absorption per reactor length unit (VRPA/H) vs. TiO_2 *catalyst load; dashed line, tubular, and solid line, CPC reactor, with 30* W/m^2 *of incident UV radiation flux. © 2010 American Chemical Society. Originally published in Colina-Márquez et al. [7] under CC BY-NC-ND 4.0 license.*

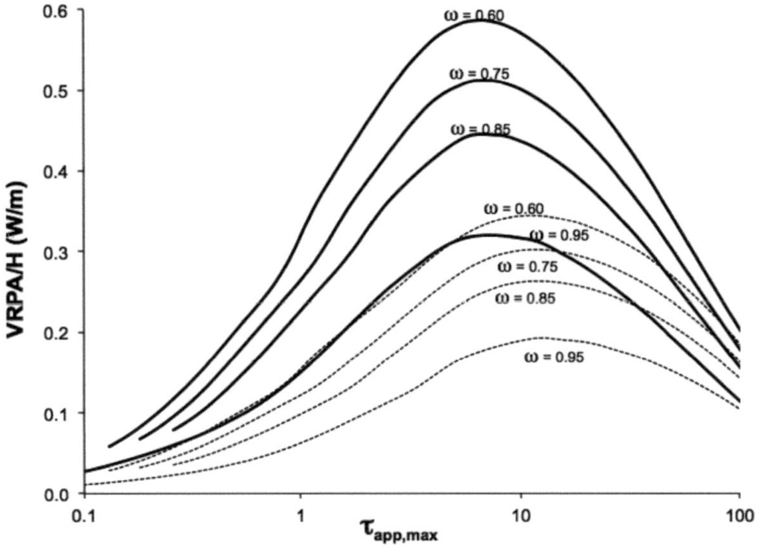

Figure 6.
Volumetric rate of photon absorption per reactor length unit (VRPA/H) vs. apparent optical thickness; dashed line, tubular and solid line, CPC reactor, with 30 W/m^2 *of incident UV radiation flux. © 2010 American Chemical Society. Originally published in Colina-Márquez et al. [7] under CC BY-NC-ND 4.0 license.*

4. Design criteria

4.1 Preliminary considerations

The design of full-scale photocatalytic plants keeps several similarities with the design of conventional plants. It is essential to consider the following parameters before starting to make the engineering calculations for the photocatalytic plant:

- Type of reactor and reactor geometry.

- Type of process regime (batch or continuous).
- Photon source (artificial or solar).
- Catalyst (supported or slurry).
- Oxygen supply (artificial or natural).
- Total handled volume.

Considering the amount of data and information gathered during the pilot-scale tests, the most relevant aspects of the plant design must be detailed in this phase.

4.1.1 Type of reactor

The selection of the photoreactor is determinant for the photocatalytic performance. The rate of photon absorption is directly related to the type of photoreactor and construction materials. The selection depends on the type of photon source and the type of catalyst as well. A reactor with a fixed catalyst can have different geometry and specifications than a slurry reactor. The same considerations apply for solar photoreactors and artificial UV-based ones.

4.1.2 Type of regime

The type of regime defines, among other parameters, the size of the photocatalytic reactor. The plant size for a batch regime is usually larger than the size of a continuous regime; nonetheless, most of the literature about photocatalytic reactors [8–10] has been focused on the batch and the semi-batch regimes. This is generally associated with the small volumes (1–100 L) handled by the pilot-scale photocatalytic reactors.

4.1.3 Photon source

The UV photon sources can be from UV-lamps (artificial) or the sun (natural). The main advantage of the UV-lamps is that they supply a constant flux of radiation, but their energy consumption supposes an additional cost for the operation of a full-scale plant. whereas, the solar radiation is a cheap and abundant form of UV photon supply (depending on the location). However, the radiation flux is highly variable, and it is only available during the daytime hours.

4.1.4 Catalyst

The fixed catalyst does not need to be separated in a later stage of the process, but it has mass transfer and photon absorption limitations; whereas the slurry catalyst can be more effective (especially for photon absorption effect), but it needs an additional separation stage that increases the costs for a full-scale plant.

4.1.5 Oxygen supply

The concentration of dissolved oxygen in water must be kept near to the saturation in order to avoid slower oxidation rates in the photocatalytic process. This oxygen can be supplied by the free contact of the falling stream with the atmospheric air, or with an air compressor with a sparging device. The free contact can be more suitable with small volumes, where the flow rate supplies enough turbulence to favor the vigorous contact between the water and the atmospheric air. Otherwise, the installation of an air compressor with the corresponding sparging system would be mandatory.

4.1.6 Total volume

One of the limitations of the heterogeneous photocatalysis, aimed for wastewater treatment, is the total volume handled by the reaction system. The larger volumes require larger installed areas for the photoreactors. Although the size of the plant depends on the kinetics of the photocatalytic reaction, the total volume determines the final dimensions of the full-scale plant.

4.2 Photoreactor selection

The criterion for selecting a reactor type considers two main aspects: performance and cost. The pilot-scale tests must provide the required information about these two features. Although we can find a wide variety of lab-scale and pilot-scale photocatalytic reactors in the scientific literature, we must ensure that the selected reactor is commercially available for full-scale applications or at least its assembling costs are affordable.

The photoreactors shown in **Figure** 7 can use artificial or solar radiation as a UV photon source, except for the annular reactor that only can use artificial UV

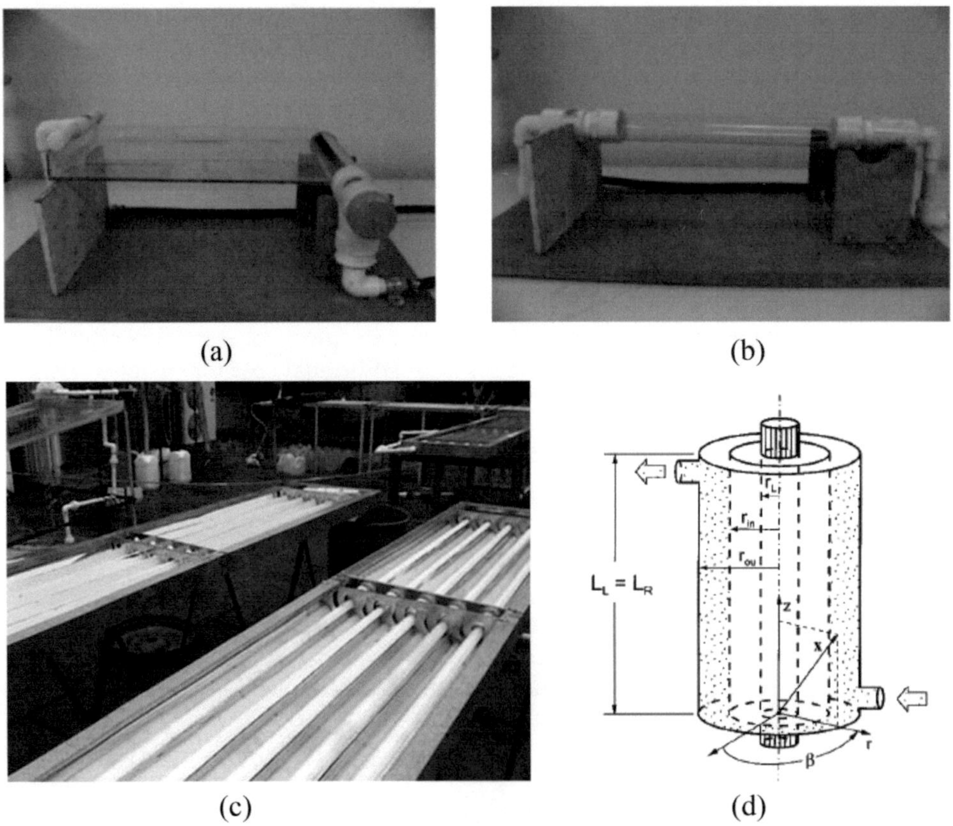

Figure 7.
Photocatalytic reactors: (a) flat plate, (b) tubular, (c) CPC, and (d) annular. © 2010 American Chemical Society. Originally published in Romero et al. [11] under CC BY-NC-ND 4.0 license.

radiation. Besides, catalysts can be used as slurry or fixed plates in any of the depicted reactors. Regarding the selection of the UV-photon source, the suggested strategy consists of decreasing costs by using solar radiation whenever is possible. If the solar radiation is feasible as a UV photon source, the photoreactor should be selected based on its performance, considering the variability of the radiation intensity.

4.2.1 Tubular reactor

Figure 7b shows a bench-scale tubular reactor. This reactor geometry is quite simple because it only consists of a tube made of a material with good UV radiation transmittance. As the catalyst is usually used as slurry, this reactor must operate under the turbulent regime in order to avoid that the catalyst precipitates.

This technology can receive direct and diffuse solar radiation on the reactor part that is directly exposed. However, some disadvantages can appear due to the soiling or the reactor wall due to the catalyst, such as mass transfer limitations and a low photon penetration inside the reactor bulk [11].

4.2.2 Compound parabolic collector (CPC)

The compound parabolic collector (CPC) consists of two reflective screens located at the bottom of the tubular reactor [12]. These reflective screens are able to reflect the radiation to the bottom of the reactor, which improves the performance of this reactor over the tubular reactor (without collectors). The CPC configuration is one of the most used in reactors for solar applications [4, 5, 13, 14]. These collectors can redistribute the radiation and reflect it without concentrate, avoiding that the fluid temperature increases [15]. This reactor uses the diffuse and direct solar radiation as a UV photon source.

Figure 8 depicts the manner how solar radiation can be redistributed around the tubular receiver surface without concentrating the radiation. The equations that describe the curvature of the involute are as follows:

$$x_{CPC} = \pm R_R(\sin t - t\cos t) \tag{3}$$

$$y_{CPC} = -R_R(\cos t + t\sin t) \tag{4}$$

where x_{CPC} and y_{CPC} are the rectangular coordinates of an involute given point, R_R is the inside reactor radius, and t is a parameter that takes values from 0 to 2π radians.

4.2.3 Tilted flat plate

The tilted flat plate reactor is also known as the falling film reactor, and it does not have concentrators. Because of its simplicity and low-cost construction, it is one

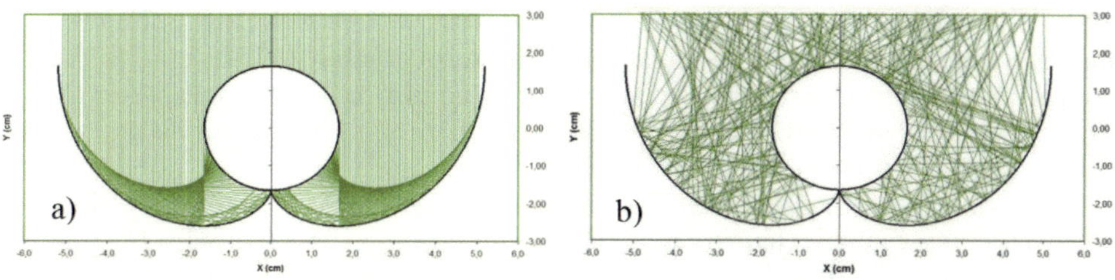

Figure 8.
Solar radiation modeled with the ray-tracing technique: (a) direct radiation and (b) diffuse radiation.

Figure 9.
Pilot-scale solar flat plate reactor used for degradation of a model pollutant.

of the most used reactors for pilot-scale studies. **Figure 9** shows a solar flat plate with a supported catalyst. The main advantage of this configuration is that the entire falling film is exposed to solar radiation with no barriers. The optical thickness is smaller than the one observed in the tubular and the CPC reactors, which permits using higher catalyst loads and achieving better photon absorption rates. However, the most significant drawback of this type of reactor is the larger area occupied compared to that used in tubular reactors. In addition, there is no possibility of setting up this photoreactor like a modular unit because it would require a pump for each module.

4.2.4 Comparison of photoreactor configurations

The selection of the most suitable photoreactor depends not only on the reactor performance but also on its potential for being scaled-up. Moreover, the type and the concentration of the substrate are determinant for the photocatalytic reactor performance. A comparative assessment can provide useful information in order to make the right decision. This means that the pilot-scale tests must be carried out under the same operating conditions, or at least under very similar ones than the expected conditions of the full-scale plant.

Previous reports about photoreactor comparisons have been carried out with artificial UV-based and UV solar-based systems [16–19]. Bandala et al. [17] study focused on the comparison of several solar tubular reactors with different types of collectors and without collectors for oxalic acid degradation. Although there was no significant difference between the performances of the photoreactors with collectors, their performance was higher than that of the photoreactor without collectors.

Figure 10 shows a case where a robust experimental design was used for showing the performance of three different solar photoreactors in the degradation of a mixture of commercial pesticides. This criterion for choosing the more suitable reactor was based on the higher signal-to-noise ratio exhibited. The signal is the overall photocatalytic removal, and the noise is the noncontrolled parameter that affects the performance significantly. This noise parameter corresponds to the accumulated UV solar radiation within a fixed time lapse.

In short, the most robust photoreactor was the flat plate, followed very closely by the CPC. This means that the selected reactor would be the flat plate for this case; however, there is a critical drawback for this reactor: a large flat plate would be needed for treating higher volumes of polluted wastewater. Therefore, the scaling-up of the flat plate would be impractical, and the best option would be then the CPC

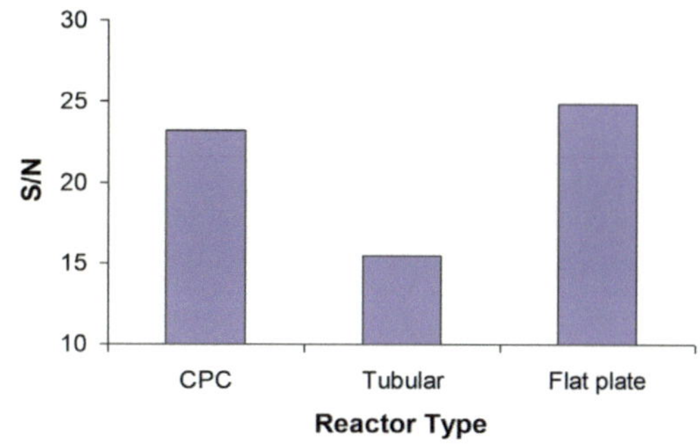

Figure 10.
Signal-to-noise ratio (S/N) for pesticides degradation with three different solar photoreactors.

reactor. Additionally, this photoreactor can be used in modules, which can be mounted or dismounted easily.

4.2.5 Selection of construction materials

The construction materials can vary depending on the reactor geometry and type. Nonetheless, all the materials considered must fulfill a common condition: they must be resistant to the atmospheric corrosion or at least, they must have protection against this phenomenon. In the case of outdoor plants, such as the solar photocatalytic reactors, the materials must resist prolonged solar radiation exposition or not undergo photodegradation under these conditions.

For photoreactors with tubular geometry, the optical properties of the construction material of the reactor must ensure the maximal use of the available photons that reach the reaction space. Regarding this aspect, the material must permit the free pass of the UV photons through it. This means that must have a high transmittance within the UV absorption range of the catalyst (e.g., 275–390 nm for the commercial P25 TiO_2).

Figure 11 shows the transmittance of different materials used in photoreactors. According to Cassano et al. [20], the material with the best transmittance is the quartz, but commercial applications with this material would be expensive. The most feasible is then the borosilicate glass with low contents of iron (Pyrex® and Duran® commercial brands). However, polytetrafluoroethylene (PTFE) can be a cheaper option with comparable transmittance respect to the borosilicate glass.

4.3 Pilot-scale plant considerations

Figure 12 shows an example of a pilot-scale plant with a TiO_2-based LFFS photocatalytic reactor aimed to degrade salicylic acid [21]. The kinetic law and the mass balance can be estimated from the concentrations measured at the sample ports and the proposed model for this photocatalytic reactor. The feed tank must be kept continuously agitated in order to maintain a uniform distribution of the concentration. Furthermore, it is essential to ensure a constant oxygen supply for avoiding that the photochemical reaction rate decreases because of the oxygen consumption. The following features must be considered for full-scale plant design:

- Type, geometry, and construction materials of the photocatalytic reactor must remain the same for a full-scale plant.

- Fluid regime (Reynolds number must be kept as constant).

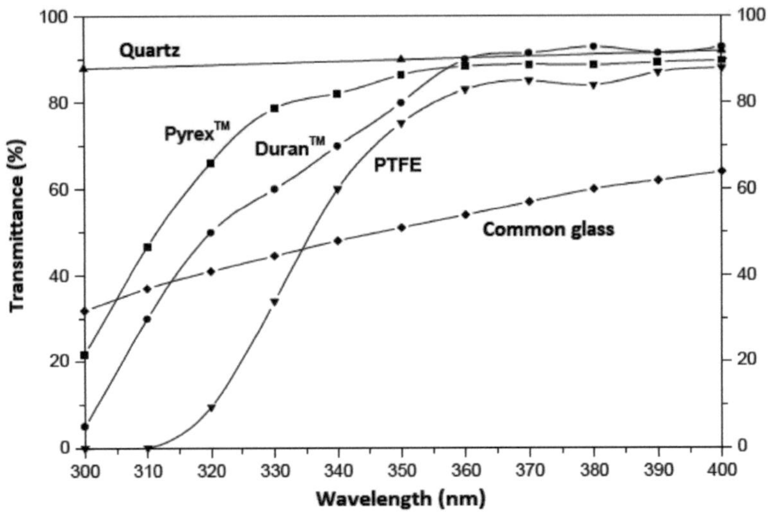

Figure 11.
Transmittance of different materials used in photoreactor design.

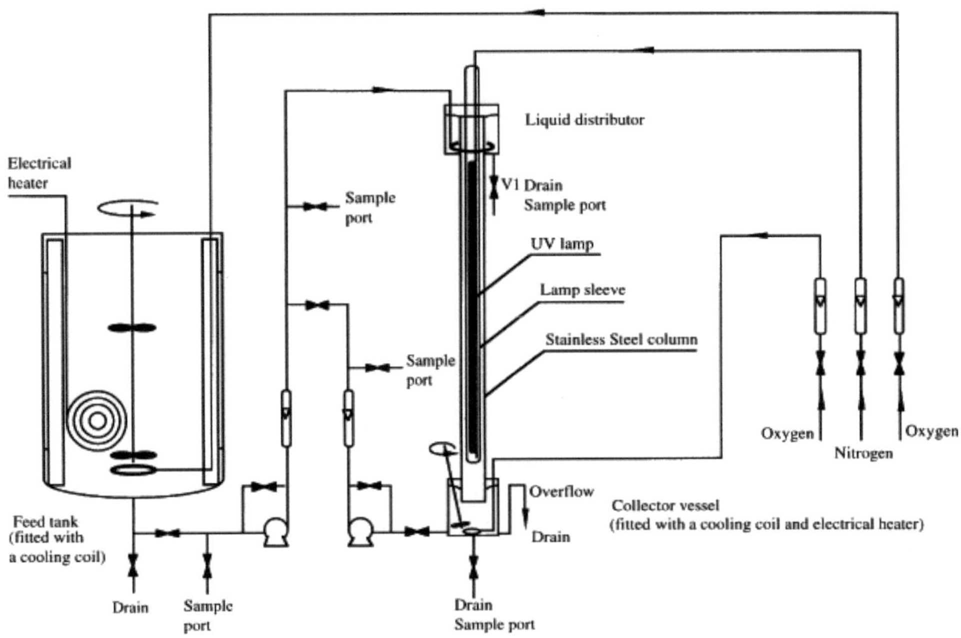

Figure 12.
Pilot plant for a laminar flow falling film slurry (LFFS) photocatalytic reactor. © 1998 Elsevier Science Ltd. Originally published in Li Puma and Yue [21] under CC BY-NC-ND 4.0 license.

- Sampling ports located at relevant streams for collecting information about substrate concentrations along the process.

- Instruments and process control for obtaining data of the operating variables such as temperature, UV radiation, pH, and dissolved oxygen.

- Easy operation.

Although the temperature may not vary significantly during the tests, it is recommendable to include cooling or heating systems for controlling the temperature, especially with systems that use UV artificial light as a photon source. Regarding the pH, it is not necessary to control it. Besides the tremendous difficulties for controlling the pH, the improvement in the photocatalytic reactor performance is not considered significant enough.

4.3.1 Plant layout

The full-scale plant can resemble the pilot-scale plant in many aspects; however, there are other stages of the process that need to be considered. Despite the process simplicity, details such as the auxiliary equipment, pumping, and storage tanks cannot be underestimated.

The photocatalytic plants are aimed usually for wastewater treatment. Thus, the final product plant (the treated water) can be reused or spilled on the surface water bodies. This aspect must be considered in the plant layout (e.g., temporal storage tanks or pipelines for conducting the treated water to its final destination). Another essential feature is the catalyst recovery. Since the catalyst is used as a slurry in most of the cases, a separation-recovery subprocess should be considered.

Figure 13 shows the isometric drawing of a demonstration plant where 2 m^3 of wastewater polluted with NBCS were treated. It is recommended to simplify the subprocess for recovering the TiO_2 because of the installation and operation costs of a full-scale plant increase significantly. It is essential to specify the pump size correctly and consider the discharge pressure in order to avoid leaks or tubes breakage. As seen in the plant layout depicted in **Figure 13**, the modules were set up in two rows of 21 reactors each. This was done to reduce the pressure at the tube's joints, but the Reynolds number must remain constant to avoid mass transfer limitations [22].

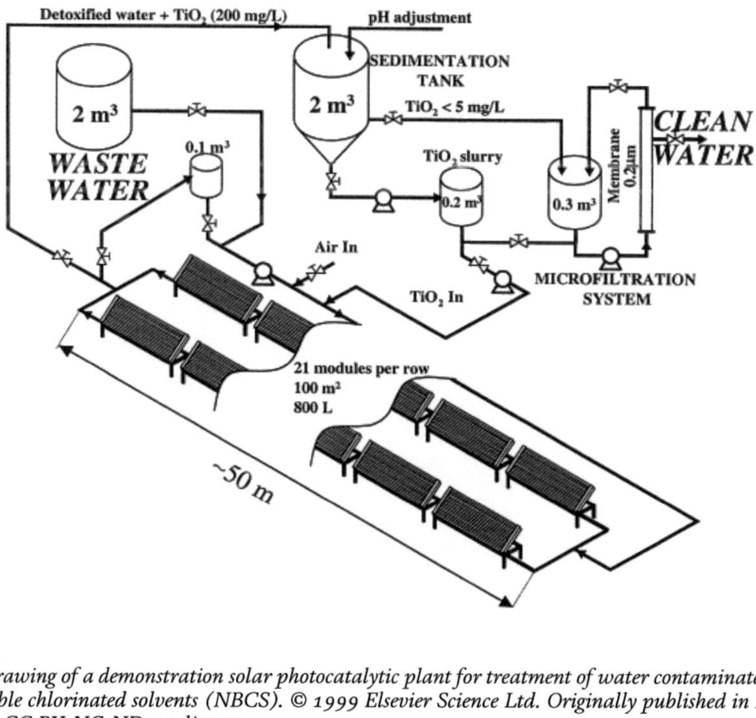

Figure 13.
Isometric drawing of a demonstration solar photocatalytic plant for treatment of water contaminated with non-biodegradable chlorinated solvents (NBCS). © 1999 Elsevier Science Ltd. Originally published in Blanco et al. [22] under CC BY-NC-ND 4.0 license.

4.4 Design, assembly, and operation of a full-scale solar photocatalytic plant

This section presents the main aspects that were considered in the construction and the operation of a full-scale solar photocatalytic plant for treating industrial wastewater from a flexographic industry. This full-scale plant is the first commercial solar photocatalytic installation in America.

Previous studies on laboratory and pilot-scale photoreactors were conducted to determine the technical feasibility of using photocatalysis as a viable treatment of 2 m^3/d of water contaminated with industrial dyes, and consequently, it served as the basis for a technology-transfer deal with the company and a patent application [23, 24].

4.4.1 Background and description of industrial process

A major company in Colombia that manufactures notebooks was facing issues with the treatment of the wastewater produced after washing the rolls used for printing the lines of the notebooks. This residual effluent was contaminated with the industrial dye, which has proven to be resistant to biological treatments. Therefore, it was necessary to apply a novel technology (such as heterogeneous photocatalysis) that could ensure the color removal.

The primary objective was the reuse of the treated water for washing the printing rolls. The most important innovation of this photocatalytic process was the simple way in which the solid catalyst is reused without using complex and expensive systems to recover after each treatment process. The use of solar radiation was another significant advantage since the location has good solar light availability.

The process flow diagram of the full-scale solar photocatalytic plant is shown in **Figure 14**. This scheme can be used for different kinds of wastewater and which can be treated by photocatalysts. This photocatalytic process was applied for treating industrial residual wastewaters contaminated with recalcitrant compounds, in this case, flexographic dye residues. The industrial wastewater is generated from washing the printing rolls soiled with dye residues after a typical work cycle. This wastewater is collected in the tank T-1 and then pumped to the recycling-feed tank

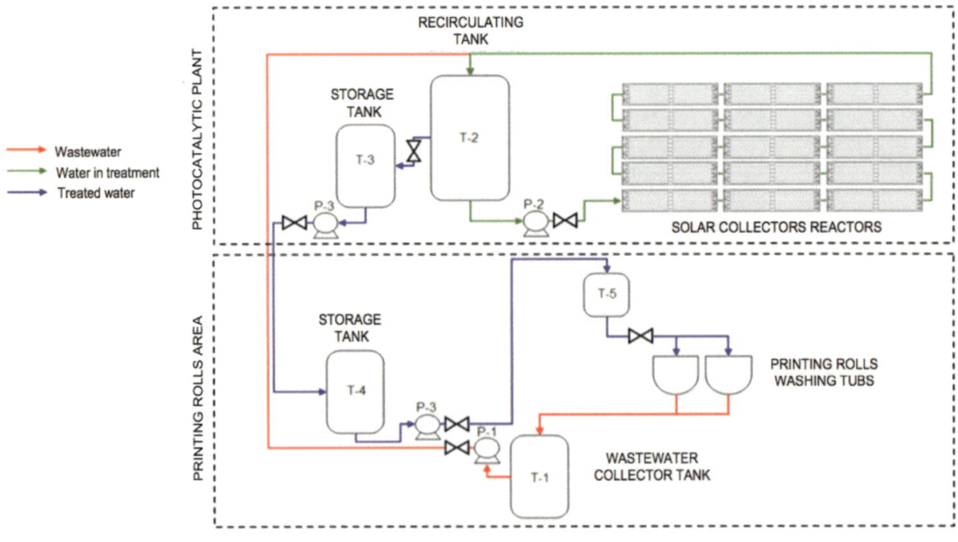

Figure 14.
Process flow diagram of the full-scale wastewater treatment photocatalytic plant.

T-2 using the pump P-1. The wastewater stored in this tank is diluted with water in a ratio of 4:1. The first dilution of the process was made with fresh water; after that, treated water is used for this purpose. Once the tank was full, the catalyst was added for the first and the only time as far. The same catalyst has been reused in the subsequent treatments, since the plant started up in September 2009, and it will be removed when it has lost its activity.

The pump P-2 recirculates the slurry through the CPC photoreactors. As it was previously mentioned, the process has an average duration because it depends on the availability of solar radiation. The process is suspended under the following conditions: color removal from the contaminated effluent, adverse weather conditions (hard rain or strong winds), or if the sun has set. In case of the color removal, the next step is the catalyst settled down in order to separate the clear supernatant. This process lasts between 12 and 24 h, depending on the amount of treated water to be reused to wash the rollers.

Between 10 and 25% of the treated water from the recirculation tank T-2 are transferred by gravity to the intermediate storage tank T-3. When this tank has enough volume (1.5–2.5 m^3), the treated water is pumped via P-3 to the secondary storage tank T-4. Finally, according to plant requirements, it is pumped via P-4 to the tank T-5, which is used to provide clean water for washing the soiled printing rolls, thus closing the treatment cycle [24].

4.4.2 Design of pilot solar photoreactor model

Each photoreactor module comprises 10 tubes placed in five rows over the CPCs (**Figure 15**). The glass tubes are 32 mm ID and 1.4 mm of thickness manufactured by Schott Duran®. The tubes were cut from their original length of 1.5 m to 1.2 m in order to correspond the length of the sheets used for constructing the collectors. The tubes diameter must be selected between 25 and 50 mm, since diameters smaller than 25 mm are not feasible because of pressure drop limitations; whereas, diameters larger than 50 mm present photon absorption issues due to a significant photon scattering.

The collectors were constructed in high-reflectivity aluminum (80–90% of reflectivity), which should be weather resistant as well. The involutes, characteristic of the CPCs, as seen in **Figure 15**, were molded using a folding machine based on

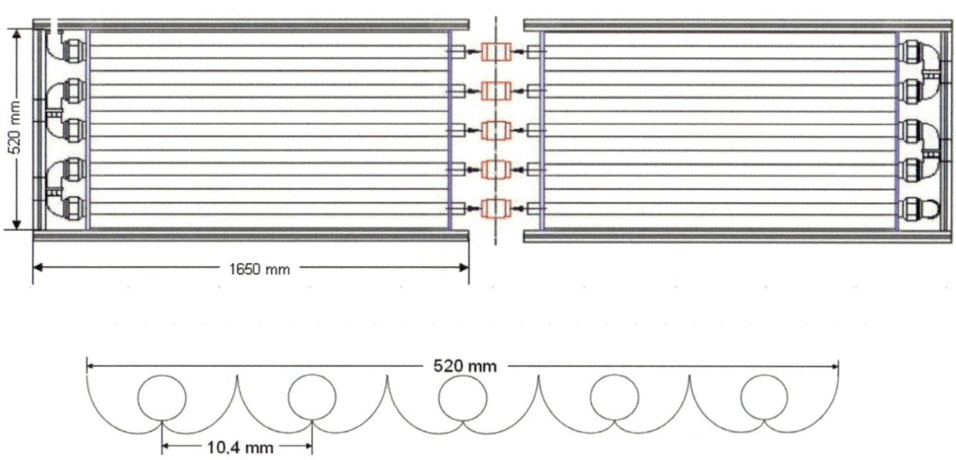

Figure 15.
Main structure of the CPC photoreactor type and cross section of tubes and manifolds for CPC.

numerical control. The supporting structure was constructed with galvanized zinc sheets, as well as the guides that supported the collectors and the glass tubes. Between the glass tubes, anti-slip PVC joints were located (red-colored joints shown in **Figure 15**), which could permit torsion without damaging the glass tube during maintenance processes.

As mentioned before, the flow regime is another crucial aspect to be considered in the pilot plant design and the scaling process. The recycling pump must be specified in order to ensure the turbulent regime of the flow through the photoreactor [23]. The Reynolds number must be larger than 15,000, and this means that the flow rate must be larger than 24 L/min (for a 32-mm tube diameter). **Table 1** shows the main accessories for hydraulic calculus.

From a simple mechanical energy balance (Bernoulli's equation), the pump theoretical power was estimated (0.241 HP or 179.5 W). The commercially available pump had a nominal power of 0.5 HP with a maximum flow rate of 35 L/min.

4.4.3 Evaluation of real wastewater at pilot solar CPC reactor

A set of pilot plant tests with a simulated mixture of effluent were performed to assess the technical feasibility of heterogeneous photocatalysis for the removal of color from industrial wastewater polluted with a printing dye. The aim was to find suitable conditions for the operation of a larger-scale plan that would allow total discoloration of the industrial effluent. For this purpose, it was considered a robust experimental design. Subsequently, a rate law was obtained, considering the UV accumulated radiation as the independent variable.

The procedure used for the experimental work consisted in treating a constant volume of the diluted mixture with different concentrations of catalyst and fixing the treatment time. The accumulated radiation was measured for sunny and cloudy days in order to observe the solar CPC photoreactor performance under these weather conditions.

The color removal was measured by UV–Vis spectrophotometry, whereby the concentration of the mixture of dyes was indirectly obtained. **Table 2** shows the conditions of the experimental pilot-scale tests.

The experimental results are shown in **Table 3**, and **Figure 16** shows samples collected before and after the photocatalytic treatment. The color was removed entirely, confirming the satisfactory performance of the solar photocatalytic reactor. This evidenced that the chromophore was eliminated. From these results, it was necessary to estimate the size of the full-scale plant for treating 2 m^3/d of wastewater. The signal-to-noise ratios for the controllable factors (dilution and catalyst load) are shown in **Table 4**.

	Quantity per meter	K_{eq}
Elbows	15	0.69
Connections	5	0.50
Valve	1	340.00
Tank outlet	1	0.50
Straight pipe	17	—

Table 1.
Fittings and straight pipe of the pilot plant.

Item	Description
Photoreactor	CPC with ten tubes Schott Duran® (32 mm ID)
Area exposed	1.25 m^2
Pump	Centrifugal pump—½ HP of nominal power
Flow rate	32 L/min
Total volume	40 L
Dye composition	Black Flexo: 50% w/w; Basonyl Blue 636®: 50% w/w
Catalyst	Aeroxide® TiO_2 P-25
Process water	Tap water
Initial pH	7.5
Temperature	25–35°C
Time of tests	10 am to 4 pm

Table 2.
Experimental conditions.

	Internal arrangement		External arrangement (Noise)	
	Design parameter		UV accumulated radiation (KJ/m^2)	
Exp	[TiO_2] (g/L)	Dilution	High (>700)	Low (<700)
1	0.4	1:8	89.3%	70.4%
2	0.8	1:8	90.0%	48.4%
3	0.4	1:4	47.4%	29.3%
4	0.8	1:4	91.5%	50.0%

Table 3.
Color removal corresponding to the pilot-scale experiments. © 2009 Walter de Gruyter & Co. Originally published in Colina-Márquez et al. [23] under CC BY-NC-ND 4.0 license.

Figure 16.
Samples collected during a typical experiment before (left) and after (right) photocatalytic treatment.

Table 5 showed the experimental data obtained for the color removal respect to the time and the accumulated UV solar radiation. This data was used for fitting parameters of a Langmuir-Hinshelwood modified model, as seen in **Table 6**.

Factor	Level	S/N ratio
Dilution	1:4	32.74
	1:8	36.59
$[TiO_2]$ g/L	0.4	33.15
	0.8	35.73

© 2009 Walter de Gruyter & Co. Originally published in Colina-Márquez et al. [23] under CC BY-NC-ND 4.0 license.

Table 4.
Signal-to-noise ratio (S/N) for the robust experimental design.

Time (min)	% Discoloration (sunny day)	Q_{uv}, KJ/m^3 (sunny day)	% Discoloration (cloudy day)	Q_{uv}, KJ/m^3 (cloudy day)
−30	0	—	—	—
0	58.1	0	0	0
20	64.5	5493.75	48.3	4481.25
40	71.0	12243.75	51.7	8953.13
60	74.2	18609.38	58.6	13312.50
80	80.7	25734.37	58.6	17690.63
100	83.9	32868.75	62.1	21881.25
120	83.9	40003.12	62.1	25218.75
180	90.4	59765.63	65.5	32681.25
240	90.4	73490.63	65.5	37762.50

Table 5.
Experimental data of the kinetic tests.

Type of day	Apparent rate constant (ppm·KJ/m^3)	Adsorption equilibrium constant (1/ppm)
Sunny	6.506×10^{-2}	3.213×10^{-4}
Cloudy	1.507×10^{-3}	9.419×10^{-3}

Table 6.
Kinetics parameters from Langmuir-Hinshelwood model.

The apparent rate constants seem to be dependent on the accumulated energy and the type of the day. This implies that the color removal will be faster on a sunny day with the same accumulated UV radiation.

Finally, for treating 2 m^3/day of contaminated wastewater with a dilution ratio of ratio 1:4, a total of 31 modules of CPC photoreactors were estimated. Nonetheless, during the project development, there were modifications to the design considerations, and 15 modules were implemented at last. An initial isometric drawing is shown in **Figure 17**.

It is important to note that a conical-bottom tank and no secondary storage tank (for clear treated wastewater) were considered at first instance. Nonetheless, the tank was sized to handle 10 m^3 per batch, and this was kept until the final design. The conical-bottom tank was selected as the first alternative in order to facilitate the catalyst precipitation and a future removal (when the catalyst was deactivated after several reuses).

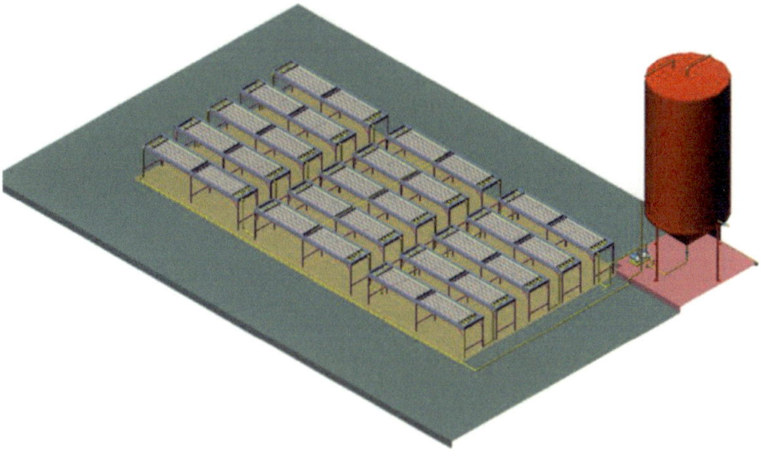

Figure 17.
Isometric draw of the solar photocatalytic plant with 15 CPC modules.

Two standard plastic tanks were selected for being installed in the solar photocatalytic plant: one of 10 m^3, designed for handling the recirculating wastewater, and another one of 4 m^3, for storing the treated wastewater. The air-sparging system in the bottom of the recirculation-feed tank was installed in order to maintain the wastewater saturated with oxygen and to avoid limitations in the photocatalytic rate due to the consumption of this chemical species. Moreover, the constant agitation of the water due to the air bubbling helped to maintain the catalyst in suspension and avoid mass transfer limitations because of the solid precipitation during the operation.

The catalyst was settled down after each operation day after turning off the air sparging system. The plant shut down at sunset and the catalyst precipitated overnight. The clear water was transferred to the secondary storage tank and later reused for washing the printing rolls. **Figure 18** shows an overview of the full-scale plant during its first day of steady operation in August 2009.

Currently, this plant is operating at full capacity (4 m^3 per day of wastewater). Some other contaminants are being fed to the photocatalytic system; therefore, it was necessary to include a pretreatment consisting of adsorption with activated carbon and a system for adding hydrogen peroxide.

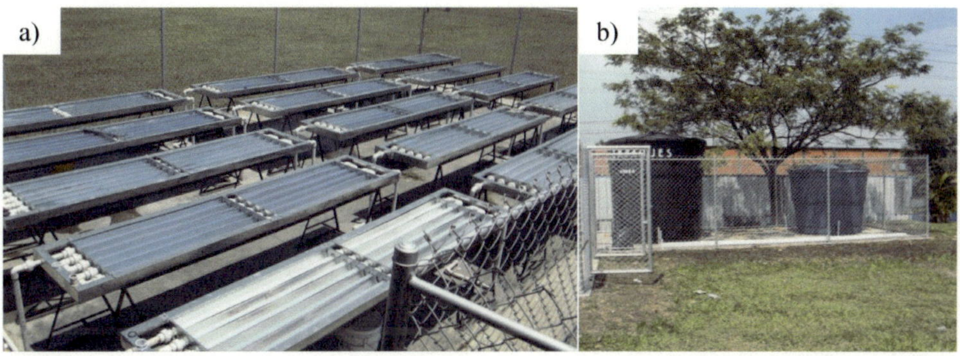

Figure 18.
Full-scale solar photocatalytic plant for dye-polluted wastewater treatment: (a) CPC photoreactors and (b) recirculation-feed and treated water storage tanks.

Author details

Gianluca Li Puma[1], Fiderman Machuca-Martínez[2], Miguel Mueses[3], José Colina-Márquez[3] and Ciro Bustillo-Lecompte[4,5]*

1 Environmental Nanocatalysis and Photoreaction Engineering, Department of Chemical Engineering, Loughborough University, Loughborough, UK

2 Chemical Engineering School, Universidad del Valle, Ciudadela Universitaria de Meléndez, Cali, Colombia

3 Chemical Engineering Department, Universidad de Cartagena, Campus de Piedra de Bolívar, Cartagena de Indias, Colombia

4 Graduate Programs in Environmental Applied Science and Management, Ryerson University, Toronto, ON, Canada

5 School of Occupational and Public Health, Ryerson University, Toronto, ON, Canada

*Address all correspondence to: ciro.lecompte@ryerson.ca

IntechOpen

© 2020 The Author(s). Licensee IntechOpen. Distributed under the terms of the Creative Commons Attribution - NonCommercial 4.0 License (https://creativecommons.org/licenses/by-nc/4.0/), which permits use, distribution and reproduction for non-commercial purposes, provided the original is properly cited.

References

[1] Li Puma G. Dimensionless analysis of photocatalytic reactors using suspended solid photocatalysts. Chemical Engineering Research and Design. 2005; **83**(7):820-826. DOI: 10.1205/cherd.04336

[2] Castilla-Caballero D, Machuca-Martínez F, Bustillo-Lecompte C, Colina-Márquez J. Photocatalytic degradation of commercial acetaminophen: Evaluation, modeling, and scaling-up of photoreactors. Catalysts. 2018;**8**(5):179. DOI: 10.3390/catal8050179

[3] Colina-Márquez J, Machuca-Martínez F, Li Puma G. Photocatalytic mineralization of commercial herbicides in a pilot-scale solar CPC reactor: Photoreactor modeling and reaction kinetics constants independent of radiation field. Environmental Science & Technology. 2009;**43**(23):8953-8960. DOI: 10.1021/es902004b

[4] Braham RJ, Harris AT. Review of major design and scale-up considerations for solar photocatalytic reactors. Industrial and Engineering Chemistry Research. 2009;**48**(19): 8890-8905. DOI: 10.1021/ie900859z

[5] Spasiano D, Marotta R, Malato S, Fernandez-Ibañez P, Di Somma I. Solar photocatalysis: Materials, reactors, some commercial, and pre-industrialized applications. A comprehensive approach. Applied Catalysis B: Environmental. 2015;**170–171**:90-123. DOI: 10.1016/j.apcatb.2014.12.050

[6] Arias F, Ortiz E, López-Vásquez A, Colina-Márquez J, Machuca F. Photocatalytic decolorization of methylene blue with two photoreactors. Journal of Advanced Oxidation Technologies. 2008;**11**(1):33-48. DOI: 10.1515/jaots-2008-0104

[7] Colina-Márquez J, Machuca F, Li Puma G. Radiation absorption and optimization of solar photocatalytic reactors for environmental applications. Environmental Science & Technology. 2010;**44**(13):5112-5120. DOI: 10.1021/es100130h

[8] De Lasa H, Serrano B, Salaices M. Novel photocatalytic reactors for water and air treatment. In: Photocatalytic Reaction Engineering. Boston, MA: Springer; 2005. DOI: 10.1007/0-387-27591-6_2

[9] Ola O, Maroto-Valer MM. Review of material design and reactor engineering on TiO_2 photocatalysis for CO_2 reduction. Journal of Photochemistry and Photobiology C: Photochemistry Reviews. 2015;**24**:16-42. DOI: 10.1016/j.jphotochemrev.2015.06.001

[10] Porta R, Benaglia M, Puglisi A. Flow chemistry: Recent developments in the synthesis of pharmaceutical products. Organic Process Research and Development. 2016;**20**(1):12-25. DOI: 10.1021/acs.oprd.5b00325

[11] Romero RL, Alfano OM, Cassano AE. Cylindrical photocatalytic reactors. Radiation absorption and scattering effects produced by suspended fine particles in an annular space. Industrial and Engineering Chemistry Research. 1997;**36**(8): 3094-3109. DOI: 10.1021/ie960664a

[12] Saravia L. Diseño gráfico de concentradores tipo CPC. Avances en Energías Renovables y Medio Ambiente. 2004;**8**(1):25-30. Available from: https://www.mendoza-conicet.gob.ar/asades/modulos/averma/trabajos/2004/2004-t003-a005.pdf

[13] Malato S, Blanco J, Fernández-Alba AR, Agüera A. Solar photocatalytic mineralization of commercial pesticides: Acrinathrin. Chemosphere. 2000;**40**(4): 403-409. DOI: 10.1016/S0045-6535(99)00267-2

[14] Malato S, Blanco J, Vidal A, Richter C. Solar photocatalytic degradation of commercial textile azo dyes: Performance of pilot plant scale thin film fixed-bed reactor. Applied Catalysis, B: Environmental. 2002;**37** (1–3):344-352. DOI: 10.1016/j.desal.2008.03.059

[15] Blanco J, Malato S, Peral J, Sánchez B, Cardona I. Diseño de reactores para fotocatálisis: Evaluación comparativa de las distintas opciones. Chapter 11. In: Blesa M, editor. Eliminación de Contaminantes por Fotocatálisis Heterogénea. CYTED: Buenos Aires (Argentina); 2001. pp. 243-266. Available from: https://www.psa.es/en/projects/solwater/files/CYTED01/17cap11.pdf

[16] Dijkstra MFJ, Buwalda H, de Jong AWF, Michorius A, Winkelman JGM, Beenackers AACM. Experimental comparison of three reactor designs for photocatalytic water purification. Chemical Engineering Science. 2001;**56**(2):547-555. DOI: 10.1016/S0009-2509(00)00259-1

[17] Bandala ER, Arancibia-Bulnes CA, Orozco SL, Estrada CA. Solar photoreactors comparison based on oxalic acid photocatalytic degradation. Solar Energy. 2004;**77**(5):503-512. DOI: 10.1016/j.solener.2004.03.021

[18] van Grieken R, Marugán J, Sordo C, Pablos C. Comparison of the photocatalytic disinfection of *E. coli* suspensions in slurry, wall and fixed-bed reactors. Catalysis Today. 2009;**144** (1–2):48-54. DOI: 10.1016/j.cattod.2008.11.017

[19] Romero V, González O, Bayarri B, Marco P, Giménez J, Esplugas S. Degradation of metoprolol by photo-Fenton: Comparison of different photoreactors performance. Chemical Engineering Journal. 2016;**283**:639-648. DOI: 10.1016/j.cej.2015.07.091

[20] Cassano AE, Alfano OM, Brandi RJ, Martín CA. Diseño de reactores para fotocatálisis: conceptos fundamentales. Chapter 10. In: Blesa M, editor. Eliminación de contaminantes por fotocatálisis heterogénea CYTED: Madrid; 2001. pp. 201-241. Available from: https://www.psa.es/en/projects/solwater/files/CYTED01/16cap10.pdf

[21] Li Puma G, Yue PL. A laminar falling film slurry photocatalytic reactor. Part II—Experimental validation of the model. Chemical Engineering Science. 1998;**53**(16): 3007-3021. DOI: 10.1016/S0009-2509(98)00119-5

[22] Blanco J, Malato S, Fernández P, Vidal A, Morales A, Trincado P, et al. Compound parabolic concentrator technology development to commercial solar detoxification applications. Solar Energy. 1999;**67**(4–6):317-330. DOI: 10.1016/S0038-092X(00)00078-5

[23] Colina-Márquez J, López-Vásquez A, Díaz D, Rendón A, Machuca-Martínez F. Photocatalytic treatment of a dye polluted industrial effluent with a solar pilot-scale CPC reactor. Journal of Advanced Oxidation Technologies. 2009;**12**(1):93-99. DOI: 10.1515/jaots-2009-0111

[24] Martinez FM, Colina-Márquez JA, Universidad del Valle. Photo-Catalysis Process Applied in Eliminating Recalcitrant Compounds in Industrial Residual Waters. U.S. Patent No. 9,394,186. Washington, DC: U.S. Patent and Trademark Office; 2016. Available from: https://patents.google.com/patent/US9394186B2/en